Caged Eagles
Downed American Fighter Pilots 1940-1945

Vern Haugland

Distributed by:
Airlife Publishing Ltd.
101 Longden Road, Shrewsbury SY3 9EB, England

FIRST EDITION
FIRST PRINTING

© 1992 by **Tess Haugland**.
TAB Books is a division of McGraw-Hill, Inc.

Printed in the United States of America. All rights reserved. The publisher takes no responsibility for the use of any of the materials or methods described in this book, nor for the products thereof.

Library of Congress Cataloging-in-Publication Data

Haugland, Vern, 1908-1984
    Caged eagles: downed American fighter pilots. 1940-45 / by Vern Haugland.
    p.   cm.
    Includes index.
    ISBN 0-8306-2146-6 (p)
    1. World War, 1939-1945—Aerial operations, American. 2. World War, 1939-1945—Aerial operations, British. 3. World War, 1939-1945—Personal narratives, American. 4. Fighter pilots—United States—Biography. 5. Great Britain, Royal Air Force—Biography. 6. Prisoners of war—United States—Biography. I. Title
D790.H383   1991
940.54'4973—dc20   91-14619
                                                                                   CIP

TAB Books offers software for sale. For information and a catalog, please contact TAB Software Department, Blue Ridge Summit, PA 17294-0850

Acquisitions Editor: Jeff Worsinger
Book Editor: Suzanne L. Cheatle
Production: Katherine G.Brown
Book Design: Jaclyn J. Boone
Cover: Holberg Design, York, PA

Contents

Foreword—
Speech by Prime Minister Thatcher,
May 12, 1986   *iv*

Acknowledgments   *vi*

Introduction   *vii*

1 The First into Action   *1*

2 The Cages of 1941   *15*

3 First to Get Home Again   *26*

4 Land of Dear Souls   *34*

5 Battles against the Japanese   *43*

6 Action and Revenge in North Africa   *52*

7 Freedom or Famine   *71*

8 The Disaster of Morlaix   *80*

9 Eagles Caged in Sagan   *95*

10 Caged in Two Camps   *113*

11 Evasions in Three Countries   *143*

12 The Last Ordeal   *161*

13 Liberation   *177*

Index   *187*

Foreword

*Speech given by Prime Minister Margaret Thatcher
at the unveiling of the Eagle Squadron Monument
in Grosvenor Square, 12 May 1986*

Today we unveil a memorial to a group of young Americans: brave men who, though their own country was not yet at war, were ready to take up arms in Britain's hour of greatest need, and that of liberty itself. Two hundred and forty-four young men; some idealists who thought not of self but who fought for a cause greater than themselves; many others, individualists imbued with a love of flying and a vision of the world they wanted to see.

Whatever their reasons, they came to this country from all over America. They came from Arizona, California, New York, Colorado, and even from as far afield as Hawaii to join 71, 121, and 133 Squadrons of the Royal Air Force. The Eagle Squadrons. How well chosen that name was! The eagle symbolized America's re-awakening from isolation. I recall the lines from Milton: "Methinks I see in my mind a noble and puissant nation, rousing herself like a strong man after sleep, and shaking her invincible locks. Methinks I see her as an eagle mewing her mighty youth and kindling her undazzled eyes at the full midday beam."

Some American pilots had already fought in the Battle of Britain, others in North Africa and Malta; all helping, in those difficult times, to turn the tide of a terrible tyranny. More followed and when America entered the war most of them transferred to the Fourth Fighter Group of the United States Eighth Air Force. I believe their new motto said that they were "Fourth but First," and in that spirit, wearing the wings and decorations of both countries, they continued in the great crusade against the inhuman dictatorship that was Hitler's Germany.

Today we remember them all, and are grateful to them as the few who were the vanguard of the many Americans who followed to fight in the air, from our shores, in the cause of freedom; a freedom which America still defends unsparingly the world over. The debt the free peoples of Europe owe to your nation, generous with its bounty, willing to share its strength, seeking to protect the weak, is incalculable. In all, over a third of the Eagle Squadrons' pilots were killed in action. From the Eighth Air Force as a whole some 28,000 American airmen were never to return. They are themselves commemorated in London at St. Paul's Cathedral in the American

Memorial Chapel below the High Altar, near to those of so many of our own great national heroes.

But today, in particular, we honour those of three Eagle Squadrons—both the living and the dead—with pride and with affection. We remember them, not simply as brothers in arms, but as true friends of freedom. The country whose cause they freely adopted as their own salutes and honours them.

Acknowledgments

The author of this book, my dear husband, Vern Haugland, worked very hard during the last years of his life to bring together these stories. An accomplished aviation journalist and writer, Vern interviewed the Eagles at length, painstakingly poured over their logs and diaries, and created a manuscript of which he was quite proud. Sadly, he died before he could see this, his final book, published.

Stories of wartime survival and hardship were particularly significant to Vern. In 1942, while an Associated Press correspondent, he was forced to bail out of a plane over the wild jungle of New Guinea. For over 40 days, he persevered against all odds until his rescue by natives. While recuperating from his ordeal, Vern was awarded the Silver Star by General Douglas MacArthur. He was the first civilian so honored. His personal diary was subsequently published as his first book, *Letter from New Guinea*.

Vern was extremely appreciative of all the Eagle pilots and their extended families for sharing their accounts of adventure. I, too, am grateful for their substantial contributions. After Vern's death, I have continued to enjoy the friendship of many of the Eagles. I want to give special thanks to Reade Tilley and Jim Gray for their support.

Above all, I want to acknowledge the love and dedication of my son-in-law, Hathaway Watson, who successfully compiled and edited Vern's work and took care of all the details leading to publication.

TESS HAUGLAND

Introduction

Half a century ago, some remarkable young Americans embarked unbidden upon a heroic journey into battle and discovery. These boys and men came from every region of America and were linked by a devotion to flying. They also cared deeply about freedom, and so they went to a war their country was avoiding, to fight among strangers in the service of a king.

In 1940 and early 1941, the cause of freedom was direly menaced. So unstoppable was the power of Nazi Germany that by mid-1940, Great Britain confronted Adolf Hitler's massive forces alone across a thin stretch of water. Britain desperately needed pilots, and in that need more than 200 Americans saw an opportunity and a challenge they could not resist. By various means they made their way to England to fly under Fighter Command of the Royal Air Force.

They did not offer their lives to this battle for money. The hazardous business of fighting the Luftwaffe offered no wealth to mercenaries. On meager wages, the Yanks flew for reasons of the spirit.

The volunteers were formed into three Eagle squadrons, whose distinguished combat records earned the respect of the British nation and their own, indeed of all the free world. Most of them went to the war somewhat inexperienced—few had been outside the United States and some had doctored their logbooks to show the required flight time—but they painfully gained the wisdom of hardened warriors. By the time they transferred to the US Army Air Forces in 1942, they were experienced, mature and skillful enough to become the foundation and leadership of the new American fighter force in Europe.

In two previous books, *The Eagle Squadrons* and *The Eagles' War*, I presented the pilots' saga essentially through their experiences in the air. Air combat is abundantly present in *Caged Eagles: Downed American Fighter Pilots, 1940—1945*, but in this book I shall play the light upon the odyssey of those Eagles who survived being downed over enemy territory to become fugitives from capture or prisoners of war in Europe, North Africa and even Asia.

Of 244 American pilots and 16 British officers who served in the three Eagle squadrons, 108 lost their lives during the war. Forty-seven other Yankee Eagles became prisoners or evaded capture after coming down in enemy territory, or were interned in neutral countries.

The Eagles were closely bound by courage and a common concern.

Many of them found that those bonds could extend to men and women in many countries, to whom they would owe their lives. In shared beliefs, hopes, and risks defied they beheld human nobility. They also came to confront the viciousness that can lie in the human heart.

In Britain, the Yanks had found a people warm, welcoming, and delightful to know. We will meet some of them in these pages. In enemy-occupied lands, men, women, and even children came forward under great risk to guide and supply the Eagles as they struggled to avoid capture. Protection was provided by the organized resistance, the Underground. Help also came to the Eagles out of simple, daring gestures by people suddenly caught up in the conflict. Even among the German military there were officers and plain soldiers who defied the Gestapo and SS to protect Allied airmen.

Among these new comrades the Eagles found courage, brotherhood, and even love. They also encountered the venality, cowardice, and callousness that led to treachery and cruelty. From within the "good farmer" could emerge monstrous betrayal; from within the calm, solicitous official could spring the sadistic inquisitor.

If caged at one of the great compounds among thousands of fellow POWs, the Eagles encountered terrible hardships. Food was often scarce; sanitary conditions, primitive. Many prisoners were severely injured; some received lifesaving treatment by German doctors, while others suffered from lack of medical care. Escape was a constant preoccupation, and the Eagles repeatedly demonstrated their resourcefulness and ingenuity in the planning and execution of numerous escape attempts. Even when caught and punished, they were soon plotting their next try for freedom.

Ultimately, the exhilarating moment of liberation would joyfully come. Having voluntarily risked their lives in a war far from home, these courageous Eagles were at long last free.

Misadventures and personal experiences of this kind are often closely held. Few of the men involved have told their stories in detail before. Now, for the first time, they speak up—in interviews and logbooks, in diaries and old letters. Their words lay bare great suffering and sorrow, as well as many acts of simple humanity and shining nobility.

Today, as they did by committing their lives to a struggle against tyranny, these men, together with their European comrades, show us by example the way of our ideals.

1
The First into Action

IN JUNE 1940, Great Britain faced great calamity, a crisis in which hope for survival was based more on desire than on power. Since the previous September, Adolf Hitler's lightning war, *Blitzkrieg*, had swallowed up Poland and western Europe with amazing speed and voracity, devouring countries in campaigns measured in mere weeks or days. Those violent fires were not burning out, but spreading. Feeding upon the Low Countries—Norway, Denmark, and France—as well as much of North Africa, Hitler clearly wanted more, especially an ignominious, terror-forced "arrangement" with Britain.

That he could not have. Instead, standing alone against him, free Britons held resolute in spite of pain against the looming horror of barbarism. The defiant words of Prime Minister Winston Churchill captured Britain's steadfast confidence:

> What has happened in France makes no difference to our actions and purpose. We have become the sole champions now in arms to defend the world cause. We shall do our best to be worthy of this high honor. We shall defend our island home, and with the British Empire we shall fight on unconquerable until the curse of Hitler is lifted from the brows of mankind. We are sure that in the end all will come right.

In the summer of 1940, all was not sure, for Hitler had resolved to invade and conquer the island. Beaches in southeast England were already targeted; barges were being gathered at ports on the English Channel to carry the troops and armor of *Blitzkrieg*.

As it had since William the Conqueror's invasion, the sea still served

England as a defensive moat. But now, the defenses of this fortress built by nature included not only the restraining, turbulent waters of the channel, but the sky above them as well.

To launch his invasion fleet free of harassment, Hitler would have to control the air above it. To drive that fleet past the Royal Navy and through the British minefields, Hitler must be able to bomb the British ships relentlessly and protect his minesweepers from aerial and naval attack. To enable his bombers and ships to strafe and shell the beaches clean of defenders and devices, Hitler would have to dominate the skies. To give his troops time and freedom to hit the beaches and begin to drive inland, he would have to enjoy a secure umbrella above them. Aerial superiority was absolutely essential to his invasion plan, *Operation Sealion.*

As they braced for the enemy's coming, the British people knew in their bones the importance of the English Channel to their survival, although most of them understood far less why the Battle of Britain would first and most essentially be an air battle. Nevertheless, they were well aware that, as their leader reminded them in early September 1940, their nation had been on this rack before:

> We must regard the next week or so as a very important period in our history. It ranks with the days when the Spanish Armada was approaching the Channel, and Drake was finishing his game of bowls; or when Nelson stood between us and Napoleon's Grand Army at Boulogne. We have read all about this in the history books; but what is happening now is on a far greater scale and of far more consequence to the life and future of the world and its civilization than those brave old days.

The Empire and the British Commonwealth of Nations bent themselves to the cause. The English-speaking peoples sent what men and help they could, and exiles from the Nazi-occupied countries joined the ranks of the defenders, but this help was at first pitifully small compared to beleaguered Britain's needs.

America was strongly influenced by isolationists, who wanted no part of Europe's "quarrels," and by well-meaning people who insisted that they were "too proud to fight." The United States maintained an official neutrality.

Still, out of many impulses—perhaps stirred by Churchill's words, perhaps moved by reports and newsreel images of the fiery devastation let loose by the Luftwaffe's Blitz on British cities, or perhaps simply aware of where good and evil lay—many Americans sought personal ways of helping. Many young Americans, eager to test their mettle, joined the Canadians and British in the struggle.

That Americans should volunteer for combat in "someone else's war" was not extremely rare. In the late 1930s, Yanks had gone to Spain as the Lincoln Brigade to fight as foot soldiers against Franco and the Fascists. In late 1939 and early 1940, some Yank pilots went to Finland to fight a losing battle against the Russians. In late 1941, American volunteers formed the

Flying Tigers to fight the Japanese over China. Others offered themselves to France, in the steps of the honored and beloved American Lafayette Escadrille of World War I.

Yet it was to Canada and England that volunteers flocked in greatest numbers. Some of them just wanted to fly—at any cost. A total of 244 were destined to join the three all-American Eagle Squadrons that were formed for the Royal Air Force. For several, especially in the early days of the gathering crusade, the going was not without its detours and disappointments.

* * * *

Eugene Quimby Tobin labeled his daybook "Memoirs of a Soldier of Fortune & a Great Lover." A bit of romantic boasting, this. From its May 10, 1940, beginning, the diary shows Tobin, a messenger at the Metro Goldwyn Mayer film studio in Hollywood, to be a rather shy 23-year-old, clean-living, friendly, religious man. Not the killer type, he went off to war in Europe because as a boy he fell in love with airplanes. Having learned to fly light aircraft, he now wanted to get his hands on one of the big birds, powerful fighters that a lowly mail clerk could only dream about, and never hope to command.

The daybook's opening passage:

> Left L.A. tonight on the 8:15 train for this "great" trip. I don't know what's going to happen, but it's sure as hell going to be an adventure....Annie, S, Suzie & Papy saw me off. Just wait till Junior finds out. I bet he has a baby—Whow.

"Papy" and "Junior" turned out to be one individual: Gene's father, I. Quimby Tobin, a Los Angeles petroleum production agent. The elder Tobin was much shorter than his six-foot-one son, hence Gene's affectionate nickname for his father. The senior Tobin told reporters years later that Gene paid more attention to mechanics and tinkering with automobiles than to his lessons at Blessed Sacrament grade school and at Loyola and Hollywood high schools.

"He was some kid," his father remarked. "His mother died when he was five. He sneaked his first ride when he was eight. I had given him $1, and he talked a barnstorming pilot into taking him up for five minutes. Even when he was just a lad he knew the sound of different airplane motors, and he could tell what type of plane was flying overhead before it landed."

High school student Tobin hung around local airports—Metropolitan, Culver City, Dycer, and others—and trained as a mechanic. Offered a part-time job, he took his pay in flying lessons. Papy Tobin recalled his son bringing home a logbook with flight time marked in "10 minutes," "4 minutes," "5 minutes." But they accumulated.

When the boy was ready to solo, he had to have his father sign a release. He brought the paper home and said, "Mind signing that, Junior?"

"I ignored the request," his father recalled. "Gene had six essays to do before he could graduate from high school. He coaxed me for days, and then I offered to make him a deal. I'd sign the release if he'd get those essays written.

"It was a deal with Gene. Everything was a 'deal.' He liked deals."

As for "Junior" having a baby—when he found out what his son was really up to, the senior Tobin was unaware of the true measure of his son's ambition to become a combat pilot. "My son went east on the eve of Germany's invasion of Holland," Tobin recalled. "He said he was going to ferry planes for an American company."

Two months and several letters later, Junior Tobin received this postcard from England: "Happy Father's Day. I'm in the Royal Air Force flying Spitfires that have a top speed of 387 mph. Address me as Pilot Officer Gene Tobin."

Actually, Gene Tobin had been making plans with a friend, Andy Mamedoff, a former barnstorming pilot who was buying a small plane on the installment plan but had been unsuccessful in his attempt to start an air charter service. Learning that Finland was seeking pilots, they had volunteered and had been accepted, at a pay of $100 a week, and had gone out and purchased cold-weather flying gear. Before the deal could go through, "something happened in Finland—there was no war there any more."

Within 48 hours Tobin was back at the studio trying to talk himself into his old job, and Mamedoff was considering the means of acquiring another plane. They inquired around and learned that an agent of the French government had been looking for pilot volunteers. A few days later, with another friend, Virgil Olson, they received day-coach tickets to Montreal and a small amount of expense money, and were on their way to Paris to join the French *Armée de l'Air*.

They had been warned to watch their step. There was concern in Washington about Americans joining the armed forces of foreign governments. The FBI was keeping close tabs over U.S. citizens leaving the country. At the U.S.-Canadian border, officers boarded the train and asked about destinations and occupations. The men said they were on the way to Montreal to visit the operator of a fish hatchery. An official opened their suitcases and rummaged through the top layers of clothing, but did not poke around. Had he done so he would have found flying helmets, goggles, and logbooks.

On the last leg to Montreal the three Americans sat in the diner and drank beer at "25 cents a crack." "It sure does cost a lot to travel, and the meals on the train—my God," Tobin confided to his diary. "Breakfast 85 cents, lunch $1, dinner $1.25. We're financing the damn war for them, that's what we're doing."

In Montreal they encountered another Paris-bound pilot, Vernon "Shorty" Keough of Brooklyn, whose specialty was parachute jumps at barnstorming shows. Pilots liked to take him aloft because, being less than five feet tall, his weight gave them no problem. The money could be good, and Keough had learned to enjoy bailing out. However, weekends, when the earnings should be best, too often were stormy. Shorty just could not make ends meet. He had accumulated 480 parachute jumps before deciding to quit. Once he jumped from 10,000 feet and delayed opening the chute until he had fallen 7,500 feet.

The train to Halifax was crowded with Canadian soldiers. "Everybody talked about the war," Tobin wrote. "They ask why America doesn't come in the war. We say, pay us the dough you owe us and we will. They look astonished."

For the journey across the Atlantic, "Andy, Shorty and Olson are on an old scow if there ever was one, loaded to the gills with mules and do they stink. Whow. I draw a swell boat, neat as a whistle, loaded with Douglas DB7s and Martin 165s. We are in a very poor position for protection, the last boat on the right hand side of the 18-ship convoy, the most valuable boat in the convoy and the one most likely to be torpedoed. The captain doesn't like it either. That makes two of us. I have lunch and dinner with the captain and engineer. They are both swell guys."

Four days later, "23 more boats join, out of Bermuda, for a total of 40, most of them loaded with gas. One incendiary bomb amidships would make the nicest fire we have ever seen. I have never seen so many boats so close together in my life, and only one light cruiser to guard the whole bunch. It's plain to see that the Allies are really hard up. The crew is very sad. They heard the German army is in France, close to Paris. They can't understand, and think there must be a traitor in one of their generals or something."

The convoy arrived in Brest on May 30, 1940, and sailed on almost immediately for St. Nazaire, 10 hours away. "So far it has been a good trip. The captain, all of the officers and the chief engineer are swell guys except they eat that damned Rocquefort cheese. I can't see how they stand it."

Further notes from Tobin's diary show the quick deterioration of the French defenses:

> 4 June. Arrived Paris, joined Andy, Shorty and Olson. All told, eight American fliers in this hotel. Germans bombed Paris yesterday, one bomb hit 450 feet from our hotel. 45 people killed within 3-block radius. All glass out of the windows around here. 10 cars hit and burned like matchwood. Holes all over place.
>
> 6 June. Passed second physical. One more big one, probably at Tours, and then I'll be a 2nd Lt. in the French Air Force.
>
> 10 June. Paris is a sad town. Everybody is evacuating. Things look dismal as hell.
>
> 11 June. To Tours. Everyone in this section is in panic. Stations at Paris and Tours jammed with people. A terrible situation; have never seen so much sadness in all my life.
>
> 14 June. Tonight when we were eating the Germans bombed us badly. The whole building shook. I thought it was going to collapse. Two planes were hit. I saw a Potez 63 chase a bomber to the ground, a good sight to see.
>
> 16 June. We left the air base, thank God. Two hours later the Germans practically leveled the place. After we crossed the Tours bridge it was

bombed and strafed. Many people caught on the bridge were killed or injured. All of them were refugees. Those lousy Germans. Went to church in Lignieres, a quaint little church. We are with the 5th Company. Slept in hayloft, ate raw meat. No food, no lodging, no nothing, dammit.

17 June. Up at 3 A.M.. Walked 11 miles before got ride to Arcay. German bombers over us all the time. Couple times we left truck for ditch; so far no hits, at least on us. Slept abandoned hayloft again.

18 June. German Dornier 15 bombed railroad 1 mile from us at 5 A.M. Old hayloft shook but didn't fall in. Walked 10 miles, left all luggage, Germans very close to us. We have to keep on the move, sleeping in fields lousy with mosquitoes.

19–20 June. Train to Bordeaux creeping along. Sleep in aisles, little food and water, filthy, morale whole train terrible. French army practically quit.

22 June. Walked to Gelaux, found rest of the boys, moved in with them. At least they have food. Heard an Armistice with Germany agreed upon. All the French officers are talking of revolt and asked us along if we want to go to England. We will go if they guarantee us flying and a commission. If not, we'll go home. To hell with the damn war.

Sunday 23 June. Forgot to go to church today but went in and prayed and told the Old Boy I was sorry. Being this is war, I think He will forgive me.

On the day the Germans marched into Paris, the four Americans found passage, along with 3,000 other refugees, aboard a British steamer from St. Jean de Luz to Plymouth. They spent four weeks first at an RAF receiving center and then at an operational training unit at Hawarden, Cheshire, where they were assigned first to the Miles Master advanced trainer. "Three times the size of anything I had yet flown, but easy to handle. The instructor said it's the closest we have to a Spitfire," Tobin recorded.

A 20-minute dual flight check consisting of "mild aerobatics" convinced the instructor his pupil was ready. "He took me to a Spitfire, two and a half tons of flying heat with a 1,075-horsepower Merlin engine," said Tobin. "He showed me how the controls work, then said to go ahead and take it up. I said a prayer, gave her the gun, and was over the edge of the field at an indicated air speed of 150 mph. By the time I had the landing gear retracted and the hatch closed, it was 170—the first time I ever had a real taste of speed. Moved along at 240, the Spit's lowest cruising speed, opened the throttle wide to make a steep climbing turn, and found I was diving at 430. Luckily I came out of the dive with a couple thousand to spare. Landing the Spit was much easier than I expected, not much tougher than an ordinary light sports plane if I watched my step."

The very next day, with only an hour and a half of flying in England to his credit, Tobin got lost on a 100-mile cross-country flight, landed at a small field for directions, went up again, and almost an hour later, lost again in clouds, tried to set down at a small field, scraped a hedge, barely missed a fence, and finally landed without damage. Armed guards surrounded the plane. Tobin was more than 200 miles off course, only 50 miles from Scotland.

Tobin's diary entry for August 14, 1940, "The adjutant called me to come to hangar 5 and take a Spit down to Hamble. I started to walk to the hangar, and was within 50 yards when I saw a Junkers 88 dive out of the clouds and drop huge bombs on the hangar. It was awful. One man had his foot blown off, another his arm up to his shoulder blade. At least three were killed, caught under the hangar door that fell. I am convinced that war is man at his worst, playing his most brutal game. I am also convinced we will win, 'cause now I'm mad."

Two days later Tobin, Mamedoff, and Keough were assigned to 609 Squadron at Warmwell, Dorset, the first American pilots to become "fully operational." Olson, meanwhile, had been transferred to another squadron. They would meet again, and fly together as Eagles.

* * * *

Tobin's logbook reflects the intensity of the Battle of Britain:

24 August. Today is Andy's birthday. He got two cannon bullets from astern shot at him. They went through his armor plating and thumped him in the back, and that is all. He is without doubt the luckiest fellow I know of.

25 August. Today at 6 P.M. had my first combat experience, probably shot down two Messerschmitt 110s. I don't know for sure because I didn't actually see them go in. I dove after one from 19,000 feet and blacked out. Came to at 1,000 feet doing better than 500 mph. 'Twas all I could do to keep the plane under control. I learned plenty about combat fighting.

5 September. Went to *New York Times* office and met a couple of American reporters, Pete Daniel and Scotty Reston, both good guys.

6 September. Went to American Embassy and saw my old friend Theo Achilles and had a few drinks. He sure is a swell guy. Met Col. Mike Scanlan, American naval attaché to London. He, too, is a swell guy. Went to RKO Studios with him and his wife (she's nice too), saw some swell shots of the raid. Had dinner with Scotty, met two more American representatives of the *New York Herald Tribune*, Frank Kelly and another.

7 September. London had its biggest raid, 500 plus planes. 400 casualties and 400 injured. Our squadron got 10 confirmed, 10 probable. No losses.

8 September. Went to church in our sleeping quarters (Church of England). Heard nice sermon.

10 September. Another big raid on London. A lot of people think the real blitz has started, so what? We'll still kick the living hell out of them.

11 September. Today will go down in history. Winston Churchill gave a speech, said the invasion is likely to start any time in the next two weeks. The Germans are really bombing London now. Every night they drop their damn bombs anywhere and run like cowards. We are bombing the Germans. Wonder who can hold out the longest? Here's a tip: England.

15 September. Toughest day yet; terrific battle over London. I shot an Me 109. Smoke came pouring out his port side. He disappeared in a cloud. Then I caught a Dornier 215, chased him, shot his aileron off, hit his glycol tank. He went in a cloud. I went down after him, saw a 215 make a crash landing near Biggin Hill. Crew of three got out, sat on the wing.

Maybe this is the ship I shot at. I'll soon find out. Jeffrey Gaunt, one of my best friends, missing. I saw a Spit during the fight going down on fire. Sure hope it wasn't Jeff. If it was, well, from now on he'll be flying in clearer skies.

17 September. Tobin, why don't you quit your beefing? You'll live through it.

19 September. Left good old Middle Wallop. We were transferred to Church Fenton today. The Sweeny American squadron is to be formed at last. Andy, Shorty and self are the seniors.

* * * *

Sir Archibald Sinclair, British Air Minister, announced in October that an all-American fighter unit, No. 71 Eagle Squadron, had been formed at Church Fenton air station, 180 miles north of London, September 19, 1940, "in the tradition of the Lafayette Escadrille of World War I." The Squadron would be composed entirely of qualified American pilots—34 of them, initially—who had volunteered for combat service with Britain. It would become a regular unit of the Royal Air Force upon completion of a shakedown period. King George VI had approved formation of the squadron and its new insignia, a pale blue embroidered Eagle with the letters ES, to be worn on the right shoulder of the powder blue RAF tunic.

The Eagle Squadron would have as its honorary commander a well-known American professional soldier, Colonel Charles Sweeny, who had been given the rank of Group Captain in the RAF Volunteer Reserve. The active commander would be another American, Squadron Leader William E. G. Taylor, a former U.S. Navy ensign and U.S. Marine first lieutenant who had joined the Royal Navy shortly before the September 1939 outbreak of war and had been flying in British aircraft carrier operations off Norway. During the early phase of operations, however, while Bill Taylor was becoming indoctrinated into his new responsibilities, actual control of the unit would rest in good part with highly experienced British Squadron Leader Walter Myers Churchill. Churchill already had received the Distinguished Service Order and Distinguished Flying Cross for his work in organizing two squadrons of Polish fighter pilots for the RAF and for his command of a squadron operating over France before the Dunkirk evacuation.

Colonel Sweeny's nephew, Robert Sweeny, 29, winner of the 1937 British amateur golf title, was named assistant adjutant of 71 Eagle Squadron.

Tobin said he, Keough, and Mamedoff left 609 Squadron with regret, "They taught us all we knew about combat and made us feel at home in England."

* * * *

The daybook for the next month reveals Tobin's anxiousness to log time on the new Squadron airplanes:

24 September. Saw our new headquarters and they look nice, but there are no airplanes or pilots yet.

28 September. Group Captain Sweeny told us we won't get started for at least three weeks. That made me madder than hell. Not a damn thing to do here but eat, sleep and fly Link trainer. Art Donahue, a new American man, arrived today. Nice guy. Was shot down while flying with 64 Squadron. Burned his right hand and face, but he's all right now. Sure wish I was back in 609 Squadron.

3 October. We are going to get Spitfires instead of Brewsters. The latter haven't arrived from America yet. Looks like at last we are going to get the ball rolling. Swell, now that winter is here and it's either foggy or hazy all the time. Couldn't have picked a better time to form a squadron. Don't worry, Tobin, you'll live through it.

9 October. S/L Taylor and Robert Sweeny arrived back from London. Rumor now is we get "Vultee Vanguard, Lockheed P38, Curtiss P40, Brewster, Bell Airacobra." I'll settle for a Taylor Cub.

24 October. Brewsters arrived. Am I happy. Tomorrow we fly them if it doesn't rain.

25 October. Flew Brewster today; damn good airplane. Has better climb and maneuverability than anything I've ever flown. Too bad it doesn't have any armament, only four guns. It would make a swell dive-bomber.

New pilots joined the squadron: Phil "Zeke" Leckrone, Luke Allen, Chesley Peterson, Arthur "Jim" Moore, Gus Daymond, Charles Bateman, Stan Kolendorski, Byron Kennerly, James McGinnis, Edwin Orbinson; and two British officers, Flight Lieutenants George Brown and Royce Wilkinson. Art Donahue, weary of the lack of action, asked for and received permission to return to his all-British squadron, 64.

28 October. Zeke Leckrone and I were flying Brewsters in formation. Came in for a landing. Zeke overshot and dumped the thing on its back. I jumped out of my plane and ran over and helped him get out. He was out colder than a clam. When he came to he didn't remember anything. Glad that outside of that he's all right. I sure like Zeke.

Soon, Tobin was to immerse himself in public relations efforts in London. "A guy by the name of Bob Low of *Liberty* magazine came out to write a story about our Squadron," Tobin recalled. "He's a nice guy, and I am going with him to London, also to make a broadcast. Gawd but this is a funny war."

Tobin went with Low to London's Lansdowne House "and met Quentin Reynolds, the writer, and a hell of a swell guy he is, too. He was soused to the gills. I like both of them very much."

Low invited Tobin, for the week's leave from squadron duty that he had been granted, to stay in the spacious Reynolds-Low apartment. They went to the British Broadcasting Corporation studio for a broadcast to America. "It was a lousy script and I'm a lousy speaker, but finally we did it," Tobin said. "I have to make another tomorrow for England. On top of that, Bob Low and self are going to grind out a story for *Liberty* magazine. God, what a screwy war."

In his 1941 book, *A London Diary*, Reynolds wrote of the breezy, good-looking flier who "eats four eggs for breakfast." Reynolds said he complained to Tobin that, "just because you have moved in, I have to love up a head waiter at the Mayfair to get you your eggs."

He said Tobin replied, "Sure, why not? Another thing, I don't drink. Be sure there's lots of Coca Cola® on hand in the icebox, and none of that tea for breakfast. I want cocoa or coffee."

Reynolds said Tobin had an answer for everything. "When a bomb fell near by, we asked why he was not up there protecting us," Reynolds recalled. "Red Tobin said, 'It's too cold to fly these nights. We only fly in nice weather.'"

At an RAF club, when his companions ordered scotch highballs, Tobin asked for a soft drink. "I say, old man, don't you drink?" one British officer asked. "I never touch liquor," Tobin replied. "I promised my father I'd never do anything unhealthy."

During his week's stay with Low and Reynolds, Tobin devoted long hours to the interviews—10 A.M. to 5 P.M. one day. "Tiring as hell." Another time, "Bob and I worked on the story till after 2 A.M." On occasions he wrote, "Came home and went to bed" while, he said, Quent, Bob, and their friends were "downstairs drinking at the bar. My oh my." Another time, "Came home, played records, sang songs; to bed at 2 A.M. This type of life has got to stop, yessir."

At the end of the week he received a check for £60 from Low before taking the train home to Church Fenton. He wrote, "Sure had a swell time with Reynolds and Low, both marvelous fellows. Best leave I ever had since I've been in England. They told me to stay with them the next time I get leave. Swell." Back at the base again, "Got a letter from my Sis Helen, also a clipping from a Denver paper saying I shot down eight Nazi planes. My Gawd. What next?"

In mid-October eye trouble sent Tobin to York Military Hospital. "They don't know what it is, and sent me to an eye specialist. He said he can fix it in a month, but I'll have to stay in London. That's going to be hard to do." At this time also the Squadron acquired its first combat-ready aircraft—nine Hawker Hurricanes—and was transferred to a somewhat larger station, Kirton-in-Lindsey, 40 miles to the southwest.

In November there was a stay at the RAF hospital at Halton for chest and head X-rays, and for a series of injections. "I'm sleeping in a 16-bed ward. The patients are mostly RAF boys, badly burned, poor guys." One of the RAF officers was a palmist: "He claims I will go through the war o.k., and make a name for myself. I think and hope he is right. Had a few drinks and came home to bed."

Tobin managed to obtain fairly frequent leave from the hospital, as he had from 71 Squadron station, to take advantage of his journalist friends' offer to stay with them in London. With his hosts he had lunch with actor David Niven, "a nice guy"; dinner at the Berkely with "a well-known English writer, Hector Bolitho, who is going to write a story about the Eagle Squadron for the *Saturday Evening Post*"; brunch at the Savoy with Harry Witt, photographer for the film *London Can Take It*; lunch with war correspondents Drew Middleton and Bob Knickerbocker; luncheon at the Savoy with 15 American reporters; dinner at the Mayfair with "Quent, Bob Low and a lot of American reporters."

He spent an enjoyable weekend at the Reading estate of "Mr. and Mrs. Miles—he is head of Miles Aircraft Company. They are both damn nice people. He showed me his new fighter—has marvelous visibility and a 1,300 hp Merlin 20 engine, two-speed blower. He invited me to fly it; damn nice considering he doesn't know anything of my flying. I flew it for 1:45. It is a damn good little trainer. He also invited me on a boat trip 'after the war' to Tahiti on his yacht. Swell, what a guy."

For a brief period Tobin won a reprieve from the hospital and was in action again:

> Flew 1:35 with Gus Daymond and Stan Kolendorski today. Daymond good kid, will be all right. Kolendorski self-centered as hell. I have my doubts. We'll see.
>
> Flew for 15 minutes today and fuel pressure dropped to ¼ pound so came home. Then I flew an hour and 50 minutes with P/O Taylor doing attacks on Andy and Daymond. Gawd, did they stink. Then I got lost again so Taylor and I landed at Finnerly and got our bearings and came home.

Tobin's off-hand appraisal of Daymond and Kolendorski was interesting. Daymond became a seven-plane ace. He rose to the rank of Squadron Leader before 71 Squadron was disbanded, and as U.S. Air Force Major took command of its successor unit, the 334th Squadron in the famed U.S. 4th Fighter Group. He survived the war and became a successful business executive. Kolendorski lost his life May 17, 1941, the first Eagle to be killed by enemy action. Of Polish descent, "Mike" Kolendorski had a hatred for the Nazis that on occasion affected his judgment, his friends said. Chesley "Pete" Peterson commented, "Mike broke formation to go after sucker bait. I had always maintained that Mike would be the first in the squadron to win the DFC, or the first to be killed."

Tobin went back to the hospital for a checkup, and wrote: "The Doc found out what's wrong with me after all the examinations—*lupus erythematosus*—whow. Started taking treatments for it today." He learned that this was a rather rare disease, a mysterious systemic fever simulating acute infection. It could develop insidiously and, because little had yet been learned about it, could be somewhat difficult to treat.

"A Jesuit padre came around," Tobin wrote. "I went to confession, first time in two months."

For a few days back at the base he was "feeling terrible—bad cold, headache." Soon he was well enough for a 48-hour leave "to see the civil Doc in London," and to enjoy another weekend with Quent Reynolds and Bob Low. He also found time for a rare date with a nurse, for a show and dinner at an inn. "She's a nice little gal," Tobin said.

Usually his diary messages of affection dealt exclusively with a girl back home. Many of the daybook entries ended with "miss my little honey," or "I sure as hell miss my little honey."

Four days before Christmas he flew for two hours "playing around and what not," and that night went to a dance in the officers' club. "Got tired and just came home and went to bed," he said. The next day he flew with five of the other Eagles to Brimcote near Coventry to pick up five Hurricanes. Then:

> 24 Dec. I have been grounded because of taking pills for my skin disease, and the phony Doc says you can't take pills and fly. He's crazy. I've been taking the pills for three weeks and still flying.
> We had a big dance tonight. Everybody soused, good time had by all. I don't like the English style of dancing. Here it is Christmas Eve, very foggy and damp, but everybody happy and drunk. Wonder how everybody is at home. Wonder what my little honey is doing, my pa and sister and all my friends. Well, 'til I see them all again. I'll say "MC"—Merry Christmas—to them from my little diary this year, but next year will be different. I'll be home again with my little honey. "MC" (everybody but Hitler, the bastard).
> 1940 has been a terrific year. Wonder what condition the world will be in this time next year. Hope this war is over. My first Christmas in a foreign country. Since I have been here I have made a lot of friends (I think). As this war took its toll, some of my pals are here no more. Yeah, they are flying in clearer skies. Great guys, every damn one of them. So boys, till we meet again.

On January 4, 1941, Tobin's twenty-fourth birthday, he went back into the military hospital, remarking in his diary: "God, is it a depressing place! Don't know what they are going to do to me up here, but I hope it's quick."

> 6 Jan. Had to take a complete physical before I could get out of here. Won't be able to fly until next Monday when I'm through taking those damn pills.
> 7 Jan. Today I learned a terrible thing. Andy told me Zeke Leckrone was killed Sunday. He hit Bud Orbison's plane at 20,000 feet, chopped his tail off, went straight in, never got out. Too bad. Zeke was a hell of a good guy, one of the best. Shorty's motor quit, he piled up, didn't get hurt, though. Went on leave today.
> 29 Jan. To London on two days leave with Bob Sweeny and Robbie Robinson, our IO. Robbie has a swell house and 20 acres in Hampstead Heath, about four miles from London. [Robbie was J. Roland Robinson, Member of Parliament on leave while serving his key post as Intelligence Officer at 71 Squadron.]
> 30 Jan. To Dorchester hotel, met Wendell Willkie and Duff Cooper, etc. Willkie seems like nice guy, but not good enough for President.
> 9 Feb. One of the best guys in the squadron was killed—Bud Orbison, what a helluva swell guy. Big, easygoing, damn good flier. Zeke Jan. 5 and now Bud, a month apart, both on a Sunday. But then Sunday is the Lord's day, when the Old Boy up yonder takes a day off. I don't know of a better day to go visiting. So to my great pals Phil and Bud, "Happy landings in your new hunting grounds."

Orbison apparently had become disoriented in thick clouds and spun in from 4,000 feet. Six days later Shorty Keough—Vernon Charles Keough—failed to return from a "scramble," a three-plane section ordered aloft after warning of a possible enemy attack. "There can be little doubt that Shorty's plane dived into the sea at a great speed and that he was killed instantly," the 71 Squadron Operations Record Book said.

A few months later, yet another of the original members of 71 Squadron, Andrew B. Mamedoff, was killed accidentally with three other Eagles—Roy Stout, Hugh McCall, and Bill White. Flying Hurricanes on a planned visit to Eglinton, Northern Ireland, they crashed into mist-hidden mountains on the Isle of Man.

An anecdote from the book *War Eagles* by James Saxon Childers relates:

> Tobin, flying over France one morning, suddenly remembered that it was Sunday and he had not been to Mass. The mere fact that there was a flock of Jerries on his tail didn't seem to matter. He was a good Catholic, and he turned head-on into the Germans, shot down one of them, and flew back to the airdrome in time to rush to the little station church before Mass was over.

Robert Low's five-part series about Tobin, "Yankee Eagle Over London," running in *Liberty* magazine from March 29 to April 26, made the flier an authentic celebrity. In his daybook Tobin wrote, "To MGM studio in Denham to see Mr. O'Brien about my story. Looks like MGM is interested in making a movie out of the damn thing. Good, and I hope they do."

Indeed, the May 1941 issue of *Metro Goldwyn Mayer Studio Club News* devoted its cover to a photograph of the former MGM employee in flying uniform and helmet, standing beside his plane. The caption said, "Leo's own Gene Tobin stepped into international fame for his valor with the RAF and his *Liberty* stories of his experiences as an American Eagle with the British."

During the same month, after the Eagle pilots complained that they were too remote from a combat area to be of much help to Britain, 71 Squadron was moved to Martlesham Heath, only 65 miles from London. In June the Americans were transferred again, to North Weald, just north of the city. For the first time, they were part of a Wing, along with two British squadrons. In August, after nine months of flying Hurricanes, they received their first Spitfires, the planes they had long been awaiting. On August 19, Virgil Olson, one of the four original Eagles, returning from a bomber escort mission over the North Sea, put out a distress call, bailed out from his damaged plane into the English Channel, and was lost.

The August 1941 issue of the same publication featured an article that said, in part:

> Officer pilot Gene Tobin is one of the most popular Americans in London. . . . Among his fellow American Eagle pilot officers, Gene is tops.

In fact, the whole squadron in a way revolves around redheaded Tobin, who leads them not only in laughter but is among the foremost in daredevil flying. I believe that if fate is kind to Gene Tobin, he will be one of America's top aces, another Eddie Rickenbacker.

Fate was not that kind to Gene Tobin. On September 7 he and eight other Eagles took part in the first sweep with the new Spitfires over France. The enemy reaction was fierce. A hundred Me 109s waited for the Spits to fly inland and then attacked. Hillard Fenlaw, a very recent bridegroom, and Gene Tobin were killed.

At Boulogne cemetery, a plain wooden cross placed over the grave gave the victim's name in French and his occupation in German:

Englischer Flieger
Lt. Eugene TOBIN
R.A.F.

For Red Tobin, like so many other young warriors, the price of glory was a cruel and early death. Yet, as one of the original Eagles, his heroism was inspirational to those who similarly risked all to help overcome the threat of tyranny. For many of them, another kind of sacrifice would be demanded.

2
The Cages of 1941

WILLIAM NICHOLS, of San Francisco, California, and Nat Maranz, of New York City, joined 71 Eagle Squadron at North Weald, England, on January 5, 1941, during funeral services for the Squadron's first fatality, "Zeke" Leckrone. Practicing close-formation flying with Edwin Orbison and Vernon Keough, Leckrone had collided with Orbison's plane and been killed. Death was quickly to become a familiar event for the Eagles, as within six weeks both Orbison and Keough were also to die in flying accidents. But other cruel ironies of fate also awaited the small group who reported that cold, harsh day in January.

As the spring and summer months of 1941 rolled past with relatively little interesting flying duty other than routine patrol, the ready-for-action Eagles became impatient, none more so than Bill Nichols. He kept signing up for *rhubarbs*—two-man low-altitude searches for enemy targets—but usually found nothing to shoot at. He told his friends at the end of August that he simply had to get into some combat immediately because, "I signed up with the RAF for only a year, and the year is up at the end of next week." The other pilots broke the news to him that he had misread terms of the contract that all the Eagles signed when they were recruited at home by the Clayton Knight Committee: they all had enrolled "for the duration of the war and *not more than a year thereafter.*"

On September 7, 1941, fate resolved both the tour-of-duty and the lack-of-action problems for Bill Nichols. In that fierce engagement with about 100 German Me 109s during a fighter sweep over France in which Hillard Fenlaw and Gene Tobin were killed, Morris Fessler narrowly escaped death after blacking out from lack of oxygen at 21,000 feet and recovering control of himself and his plane only very close to the ground, and Chesley Peterson shot down his first enemy plane. Bill Nichols bailed out of his burning Spitfire into German-occupied territory and became an early prisoner of war—though not the first to be so caged among those who flew as Eagles.

That lamentable distinction belonged to Nichols' compatriot Nat Maranz, who flew for five months as a member of 71 Squadron and then was attached briefly to the Second Eagle Squadron, No. 121. On June 9, 1941, Maranz transferred to RAF No. 1 Squadron. Thirteen days later, he was shot down and captured—leading in longevity an honor roll of twenty-six brave Eagle pilots to be imprisoned at Stalag Luft III, Sagan, Germany.

* * * *

The first pilot to become a POW while still an Eagle Squadron member was William I. Hall of 71 Squadron, who bailed out of his Hurricane Mark II near Lille, France, July 7, 1941. Hall said that while "sleeping on and off and waiting for the scramble," in the squadron dispersal hut at Martlesham Heath, 65 miles northeast of London, that day, he had a startlingly prophetic dream.

> The gong for scramble wakened me out of a nightmare in a cold sweat. I had dreamed that I had been shot up and wounded on my right side and that my right side was paralyzed. The dream was so real that I mentioned it to Robbie, our intelligence officer, en route to scramble.
> Our mission was to escort bombers on a daylight raid over Lille. We were in heavy flak the minute we crossed the coast, and just before arriving at Lille I was hit on the right wing by ack-ack. Part of my wing was destroyed. We were at about 10,000 feet. I started to lose altitude and spun out. At about 1,500 feet I got straightened out and started to follow the Squadron at low altitude when a Jerry hopped on my tail. He got several bursts of .30 calibre into my tail, and hit the armor plating of the back of my seat. It sounded like someone playing the traps.
> The Jerry was also firing a cannon. Luckily for me this did not hit the back of my seat, or I'd have been a dead duck. Instead, it came up through the cockpit and into my reserve tank, dumped gasoline into my lap and set the plane on fire. When the shell came up through the aircraft it got me in the right knee and paralyzed my right side. All I could do was to roll her over, open the canopy and drop out.
> I didn't take the time to disconnect my helmet with its radio cord and oxygen line so, as I went out, of course I got wrapped around the stabilizer on the tail. Then I tried to open my parachute with my right hand. I couldn't do it, so I had to use both hands, using the thumb of my left hand to push my right hand to pull the rip cord.
> When my chute opened I could see the leaves on the trees. I swung once in one direction, once in the other, and hit the ground. I landed right behind a hedge in a dazed condition, and noticed people hoeing in the field. They wanted me to get up and run, but I could not move. Almost immediately some German soldiers with fixed bayonets came through the hedge. Someone went to get a stretcher, and then they carried me back to a German officers' mess hall about 50 yards on the other side of the hedge.

Hall lay on the stretcher in the hallway of the mess from about 1 P.M. until 9 P.M., when an ambulance arrived.

> Of course, all the Germans came in and took a good look and talked. Some spoke English and some didn't. They poured quite a bit of brandy into me as I lay there in not too good shape, and they gave me a certain amount of medical assistance. My kneecap had been shot off, and I had shrapnel in my right side.

As Bill Hall lay badly wounded, he had no way of knowing the July 2, 1941, Lille mission was being celebrated back at Martlesham Heath. Three of his 71 Squadron mates had scored the first air victories yet chalked up by the Eagles. Henry "Paddy" Woodhouse, British leader of the squadron, and Americans William R. Dunn and Gregory "Gus" Daymond each shot down a Messerschmitt 109, and Bob Mannix was credited with one probably destroyed. Of the 12 squadron pilots on the sortie over France, Hall was the only one who had failed to return safely.

The ambulance took Hall to a hospital in the village of St. Omer—"a ride of about three hours because we went around to different spots to pick up German wounded. They operated on me that night." One somewhat strange incident caused the American to worry over the nature of the surgery about to be performed.

> Just before I went under the ether the doctor, with a couple of nurses looking on, pulled up my hospital gown, rapped me on the nuts with his scalpel and said, "You won't be needing these any more, will you?" I replied, "No, not for a while," and he laughed. The next morning that was the first thing I checked, and everything was okay. He was a real good doctor and I thank him today for saving my leg. He operated on my leg nine different times.

Content with the conscientious medical attention he was getting, Hall was able to share brief human bonds with a German foe:

> At one time I was sitting in line waiting for the operating room. They were pretty busy that morning and they brought in a German flier. He had been burned very badly. He had been shot up and had to bail out of his Me 109 that was on fire. The skin was hanging off the ends of his fingers, and his face was badly burned.
> When the orderlies came to put me in the operating room, I asked them to take him first because he was in such bad shape, and they did. Later on—I guess he was there about a month, and it was probably about the middle of August—his mother brought him a bottle of anisette for his birthday. He came upstairs with his mother and he poured me one or two glasses and we had some birthday cake. His mother thanked me. He came to say goodbye when he left the hospital, and that was the last I ever saw of him.

For the first three or four days at St. Omer hospital, the armed guards outside Hall's second-floor room would play two phonograph records, "Roll Out the Barrel" and "Hanging Out the Washing on the Ziegfried Line," over and over, steadily, hour after hour. "Each time they got a new prisoner they would do the same thing to him, not to please but to intimidate or annoy." Also occupying beds in the room were a badly burned Polish flier, still in critical condition, and an RAF man, Tommy Harrison, who had been shot in the leg but was getting along quite well.

Hall continued:

> Eventually Harrison and I planned our escape. A little French ward maid had brought me a small surgical saw and I had my cast slit all the way down and we practiced walking at night. The doctor decided to operate on my leg again, as gangrene had set in. After the operation he told me he had never had such an easy time removing a cast.

Escape, however, continued to be an illusion for the wounded flier, who instead became exposed to primitive, yet effective, healing methods:

> Once again I was tight into a body cast, and in traction. The doctor was fighting the gangrene in my leg and had opened the cast at the knee. During the day I was placed in the sunny window so the flies would land on my gangrenous knee. The flies landed, laid their eggs, and the maggots ate the dead flesh away. Eventually it cured my leg. The German doctor said there had been so many wars fought in France that the earth was full of gangrene bacteria.

Five weeks after Hall arrived in the St. Omer hospital, he and his two friends acquired another roommate, a man already famous around the world, Wing Commander Douglas Bader, Britain's legless fighter pilot. Bader had lost both legs in a 1931 crash of a small fighter plane, the right leg above the knee, the left leg six inches below the knee. In 1939, with a war on, the RAF permitted him to take flight training once again, with the aid of artificial legs, and in January 1940 he qualified to fly the Hurricane. By the end of July 1941 this remarkable man was the RAF's fifth ranking ace with 20½ victory credits. On August 9, 1941, leading his Spitfires on bomber escort to targets at Lille, Bader shot down two more Me 109s only to have his own plane torn in two. He parachuted to earth without his right artificial leg, but with the left one still in place.

"There was great excitement among the Germans over having Bader as a prisoner," Hall said. "The German officers from the squadron based close by came in to see him every evening. The Germans found his right leg jammed in the wreckage of his plane and brought it to him in the hospital. It was twisted all to hell, so they took it away and repaired it for him. Bader persuaded the Germans, however, to put out a request on the radio to the British to drop a new leg for him. The RAF later dropped the leg over St. Omer en route to a bombing mission."

While Bader was inspirational and influential enough as a pilot, Hall was glad that the German doctor declined to follow his medical opinion:

> Bader was in the hospital operating room one day and watched the procedure the German doctor was following in the treatment for gangrene. Having done so well with artificial legs, Bader advised the doctor to amputate my leg. The doctor replied, "No, my father was a doctor in the first war and he amputated a good many legs. After the war you saw a number of Germans walking around with wooden legs." The doctor said if there was any way possible, he would save my leg. He told me, "I am going to do the best I can for you." And he did.

The German nurse, Sister Erica, was always very kind to the patients, Hall recounted. Although she could speak only a few words of English, she expressed herself through small kindnesses—occasionally supplying him a lemon, sometimes a couple of cigarettes, or at odd times candy, or by sewing a button on Bader's tunic.

> There was also a little French girl who worked as a ward maid and had connections with the French Underground. When Bader planned to escape, she plugged the sink in our room and a French plumber, who was part of the Underground, was called in to repair it. What transpired from that meeting I never knew. I was set for another operation, so was out of it.
>
> It must have been on the night of September 9 that Bader decided to make his escape. I tied several bedsheets together for him with square knots, and he secured this to a bedstead. Some time between midnight and 1 A.M.—it would have been between guard changes—Bader threw his rope of sheets out the window and told me there was enough and some to spare.
>
> It was a very calm night. The courtyard below was all cobblestones. I heard Bader slide down a sheet and stop at a knot, and slide down another sheet and stop again at a knot until he reached the ground. In the quiet night, every step he took, his legs squeaked like the devil. I could hear him squeaking off up the cobblestones and making quite a racket. How he got away with it I'll never know.

(What Hall did not know was that the Underground had arranged for a young Frenchman to wait outside the St. Omer hospital entry gate for two hours each night, in case Bader should manage to get out. The young man escorted Bader to a hiding place, but late the next day German house-to-house searchers found it and recaptured the legless wonder. Bader made several other escape attempts, including tunneling efforts, in the various camps to which he was transferred. Bill Hall was to see him but once again, at the gates of Stalag Luft III in January 1943 as the indomitable Bader was being led off to a maximum-security camp.)

What was apparent the next morning, when the Germans found Bader was gone, was that all hell was breaking loose. Harrison and the Polish flier were shipped off to Frankfurt am Main. Hall was rushed to the operating room to have his body cast changed to a traveling cast, and the following day he was shipped to a German hospital, staffed by British doctors, at Holmark, Belgium, on the German border.

Conditions at Holmark were hardly favorable. "The doctors were very good but they lacked medical equipment. They couldn't do much. They had me in a traveling cast, extending from my hip down to my feet. There were many RAF patients—gunners and others—who had all been shot up quite badly, and there were no medical supplies for them. One day a doctor came in to bathe a patient in the antiseptic disinfectant Dettol, and someone had mixed up some Lemon Jello and put it on the shelf. The doctor mistook this for Dettol and bathed the lad in it. Before this the wounds had resisted treatment, but after the Jello they started to heal. From then on, the doctors used Lemon Jello to heal wounds."

Hall's next stop was a five-day stay at the Luftwaffe interrogation center at Frankfurt am Main. On the way from Holmark, at night, the train stopped at Cologne. A raid by RAF bombers was on, and everyone had to leave the train and go to tunnels under the station. Hall was on crutches, unable to negotiate the stairs, but his guard got a stretcher for him. There was quite a crowd rushing down the stairs. As two German soldiers carried him down on the stretcher, somebody slapped him on the shoulder and slipped a package of Camel cigarettes underneath his head, all at the same moment. Said Hall, "I never even saw the man. This act of kindness told me that somebody was right there with us practically all the time."

> At Frankfurt they asked me all sorts of questions. The interrogator showed me a copy of *Life* magazine only about 10 days old with pictures of almost our whole Squadron. The German had all the data. He knew more about the situation than I did. I spent the next month, October 1941, at a POW hospital at Statroda staffed by British doctors who had been picked up at Dunkirk—not a bad hospital, but very crowded. Then they moved me to Hildburghausen POW hospital in the heart of Germany where a British army surgeon, Col. Wilson, was in charge. The hospital was encircled by two barbed wire fences about 15 feet high with barbed wire coiled between them, and was patrolled by armed guards. I was there for more than eight months.
>
> The hospital routine was the same every day. In the morning, when you could move, you'd get up. When you couldn't, you'd just lie in bed. The time was just from one day to the next, and you never knew where they were going to shove you.
>
> One of the patients, a Sergeant Major McCarthy who had been a POW in World War I, taught me to knit. We unravelled Red Cross socks and used coat hangers for knitting needles. The hospital was always cold—there was no heat—and I knitted myself a pullover sweater that I wore through the rest of prison camp and for many years after the war.
>
> At Hildburghausen we read in the Berlin paper that a German baker had been awarded the Iron Cross for making bread of 70 per cent sawdust and 30 per cent flour. That explained the horrible black bread. Shortly after I arrived there the Red Cross arranged through the Geneva Convention for a small monthly payment to each POW. This was made in POW reichs-

marks, and apart from being able to buy toothpaste from the German PX with them, they were useless.

They kept piling up until we had great wads of them and only used them for playing poker. Finally through the German sergeant, Willie the Feldwebel, the merchants in town were persuaded to accept these Gefangenen reichsmarks, which could be exchanged in Switzerland for German reichsmarks. Willie took one of the fellows in to the town and he did the purchasing. I don't know how much money the chap had, but it was in the thousands of reichsmarks. He bought up everything he saw in the store, from trumpets to flutes.

He tried to buy a piano but didn't have enough money. It came to the point finally where the townspeople complained about us cleaning out the stores, and the merchants refused to sell to us. So that was the end of that. I remember a beautiful ebony clarinet with ivory and silver keys. I learned to play "Nearer My God to Thee," and that was all.

Willie the Feldwebel was always after chocolate from our Red Cross parcels, and for a chocolate bar he would do almost anything. We talked him into taking us for walks, maybe half a dozen of us, most on crutches, and we'd walk around the back lanes of the town. More than once Willie went into the local pub and brought us out a pint of beer to be drunk in the back lane. One day Willie wanted to show us his car and took us to town and in to a garage, and there was a VW bug, probably built in 1939 or before, stored for the duration of the war.

Once or twice we talked Willie into getting us a keg of beer for the hospital on a Saturday night, but inevitably Willie's superiors would hear about these things and they would end. I finally got out of my cast and was to be shipped out in about a week's time. However, I had ended up with a stiff leg, and the doctor planned to put me under anesthetic and try to bend my knee before I left. The night before the operation Willie got us a keg of beer and four of us attempted to drink the keg. The result was that I fell down the wide marble stairway on my crutches, ended up at the bottom, and when I came to my knee was bent around behind me. No harm done, but from then on I could bend my knee.

* * * *

While Bill Hall was being shuttled between medical facilities and recuperating as best he could, his fellow Eagles were carrying the fight to Nazi-occupied France. And as 1941 wore on, three more young men of 71 Squadron fell into enemy hands.

Bill Geiger—an athlete, the captain of his fencing and ski teams in college in California—was 12 days short of his twenty-second birthday when he was shot down off Dunkirk September 17, 1941, while taking part in escort of Blenheims on the largest RAF daylight bombing raid yet launched over France. He barely managed to bail out of his Spitfire, and was picked up by a German E-boat a few hours later from his dinghy in the English Channel. One of his good friends in 71 Squadron, 18-year-old Tom McGerty of Los Angeles, was killed.

Upon his arrival at Dulag Luft, the Luftwaffe intelligence center at

Frankfurt, Geiger experienced the boredom and solitude that can overwhelm a young and vital warrior taken out of action.

> I was put in a room—a comfortable enough room with a bed, table, chair—but I was by myself. There were no magazines, nothing to read, nothing to do, and I was left alone there for six or seven days. We had not been briefed on what to expect as a POW. I thought this might be my life for the rest of the war.
> Finally a man by the name of Eberhardt came in with a bunch of magazines and a lot of conversation, and I was so glad to talk to somebody that I am sure I probably said a great many things more than my name, rank, and serial number.
> Fortunately I did not have any particular information, and I probably could not have given him any information that was of any value to them. But the trick worked pretty well. You are a young kid, you have just been through being shot down and fished out of the English Channel, and as a fighter pilot you are alone. You spend seven days talking to no one but yourself. Then all of a sudden you have an opportunity to talk to somebody who has a lot of charm and doesn't seem to be asking any particular questions, and the tendency is to run off at the mouth pretty freely.
> Toward this end, I might say as an ex-POW that I think the books should be rewritten on what you can do and what you cannot do. I think prisoners of war should be allowed to sign anything they want to sign, say anything they want to say, that will keep them from being tortured or mistreated. I think we should publicize this so that the whole world knows that our prisoners of war will say whatever it is they are asked to say. In this way it will be of no value to the enemy. There isn't any point in going through torture unnecessarily. The best policy is to make sure that the people flying combat missions don't have secret information, and then let them say whatever they want to say if they are shot down and captured.

Bill Geiger was soon to be reacquainted with the next Eagle to be caged. Gilmore "Danny" Daniel, an Osage Indian from Oklahoma, was the youngest American flier—and possibly the youngest fighter pilot of any of the Allied countries—to be shot down by the Germans in World War II. Danny Daniel was only 15 when an Me 109F got to him during a bomber escort mission to Bethune, France, on October 13 1941. Upon enlisting in the RCAF in December 1939, Danny, whose Indian name was *Ku-sha-he* (meaning "Little Two Shoes"), presented papers showing a birth date of November 30, 1921. "Back when I was born, records on the Indian Reservation were not properly kept," Danny confessed years later. "My father got a doctored birth certificate for me. It was not until long after the war, and the Indian Agency had searched many family histories and records, that I learned my correct birth date was November 30, 1925."

Enrollment in the Oklahoma Military Academy at Claremore at the age of eight helped Danny get an early start in aviation. He learned to fly at the Spartan School of Aeronautics in Tulsa "as a sort of extension of the military academy." On his way to becoming a fighter pilot, Daniel trained at an

RCAF Instructors' School at Moose Jaw, Saskatchewan, run by the brothers of Bill Hall, whom Danny was to know like a brother later at a POW camp. The two had not met as Eagles, since Hall was shot down almost three months before Daniel joined 71 Squadron.

After Daniel parachuted into the Strait of Dover about five miles off the coast of England on October 13, 1941, he drifted in his dinghy for 78 hours, only to be washed ashore in France and taken by the Germans to a hospital at St. Omer, the same hospital where Hall had been a patient and from which Douglas Bader had escaped.

In a narrative in *The Eagle Squadrons*, Daniel recalled:

> I was placed in a room with two other pilots, a New Zealander and a Rhodesian. Both were badly shot up and burned. As they were sergeants and I was an officer, the Germans always talked to me first and I relayed everything to them. After two days we were removed to Lille Hospital for a week because typhus had broken out at the St. Omer hospital.
>
> In the hospital I had a visit from a Hauptmann, or Captain, Phillip. He told me through an interpreter that I was his forty-eighth victory. I followed his career in German newspapers all the time I was a prisoner of war. He was killed on the Russian front three years later, and the Germans said he had 350 victories. At Lille I was interrogated by an intelligence officer who wanted to know what I was flying when I was shot down. I told him to go ask Hauptmann Phillip—he knew. The interrogator got so angry he slapped me and said I was an American mercenary and would be shot.

Daniel was transferred from the hospital to Dulag Luft for interrogation. After four days he was put into the receiving compound, where Bill Geiger and Bill Nichols met him at the gate. A few days later Morris Fessler arrived.

During an October 27, 1941, dawn attack with Wally Tribken against a freight train engine in Boulogne, Fessler's plane was damaged, and he was forced to land in a plowed field. He set his plane afire, eluded searchers all day, and late at night was admitted into a farmhouse. Two French police officers were among the persons there.

Fessler carried an English-French pocket dictionary, had learned some French in school, and thus could communicate with his French hosts. They offered him soup and wine, but said they feared harsh punishment from the Germans, who would certainly be coming around, if they should learn of aid offered to an American pilot.

"I knew they were telling the truth," Fessler said. "I wondered whether the gendarme with the pistol would fire at me if I were to run out the door into the night. At the same time, I wanted none of these people killed because of me. Therefore, I allowed the gendarmes to take me with them and turn me over to the German authorities. Thus began my three and one-half years as a prisoner of war."

For the next six weeks, the Eagles at Dulag Luft led a comfortable life. America was not at war yet, and the German captors tried to encourage a benign approach to world events. As Danny Daniel reported: "While we were at Dulag Luft the Germans encouraged us to write letters to the U.S.A., as many as we wanted to, to let the Americans know we were o.k. and it would be a mistake to aid the Allies in the war. We were taken on walks in the countryside and stopped at taverns, and we also were told we would be repatriated as soon as it could be arranged. Of course, the Pearl Harbor attack of December 7, 1941, ended that."

For the first wave of caged Eagles, 1941 thus proved to be a year of disappointment, hardship, and self-confrontation. Yet, despite their uncertainties over the length and nature of their confinement, these young warriors-turned-men already had begun to display the sense of humor and optimism that would carry them as long as it would take them to regain their freedom. In late 1941, however, they could only anticipate a future of more camps and, inevitably, reunions with their fellow Eagles.

* * * *

At the end of the year, one Eagle found himself in confinement, not in the hands of the Germans or their allies, but in neutral, fiercely independent Eire.

Roland E. "Bud" Wolfe of Lincoln, Nebraska, had joined 133 Squadron in England in August 1941, the month it was established. In early October the Squadron was ordered to Northern Ireland for advanced flight training. On the way to Eglinton four of the pilots—Andrew Mamedoff, Hugh McCall, Roy Stout, and William J. White—flew into storm-shrouded mountains on the Isle of Man and were killed. Wolfe's brush with death occurred November 30.

Returning from a convoy patrol north of Ireland, his Spitfire engine quit. He had lost the coolant and had very little time to fly. He had an engine temperature instrument and could keep the engine going for a few minutes also by using the electric primer. He was very low on fuel, on top of an overcast, and his radio receiver had failed. Unable to get help from ground control, he was struggling along a dead-reckoning course that he hoped would lead him to the 133 Eagle Squadron temporary base at Eglinton just outside Londonderry, Northern Ireland.

Wolfe gave thought to this predicament and bailed out. "No one could survive long in the sea north of Ireland. I headed for the closest land. As our Squadron Leader George Brown used to say, 'When the urge box ceases to urge, you nip smartly over the side.' So I did. I saw my flying boots take off in formation as the chute came full open. I never did find them. I came down in rough country in County Donegal but was not hurt much."

By bad luck, Wolfe had come down only about 10 miles from his destination, but on the wrong side of the border. He had descended into the Muckish Mountains of Donegal, in the northern part of Eire. The local police found him in short order and put him in jail until they could telephone the Army.

Wolfe recalled his extended sojourn in Irish internment:

The Irish were friendly enough—many Free State men were in the RAF. However, the Government policy was not so friendly. In fact, they locked up all who came down in their country, and they also locked up most anyone who opposed them.

From Donegal, the northernmost county in Ireland, I was taken out around Ulster and south to County Kildare. There were no other Americans in this internment camp—just English, Canadians, one Free French, two Poles, one New Zealander and myself. All of the internees were aircraft crew members, both officers and enlisted men.

We had a radio and could buy a local newspaper, so we kept up to date. We used to stay up late to get the BBC news which, in daytime, was drowned out by static from the local Irish station.

In an internment camp you don't look for much in the way of housing. We used peat for heating and cooking, and peat gives off little heat. We obtained food from the basic army ration. No one gained weight. Sometimes we could buy extra food, but as in England, things were in short supply. The response to requests would be, "There is a war on."

About mail: we got it. It took some time, but most letters came through. Packages were something else. Only one from home reached me, and I wrote my family to stop sending them. Everything was subject to censorship, of course, and some people can be tempted only so far.

We saw little evidence of the Catholic-Protestant hatreds within Ireland that are so much in the news, year after year. Our camp was far to the south, where Protestants were few in number. They sat and kept still, I guess. We did not get to know many civilians.

We saw very little of the German prisoners. Their camp was adjacent to ours, but there was a 10-foot tin fence in between. There was much barbed wire all around and about, and guards, lights—the whole ball of wax.

About my escape attempts? You can go over, under or through walls, barbed wire and other barriers. That's all there is, and we did them all.

Yes, I got mauled about somewhat several times, but only because of the escape tries. The other side did not get off without a few bruises. All prisoners were treated alike. Fair enough if you would stay put. Some would not. I'm sure the German prisoners got the same treatment.

In fact, Bud Wolfe managed to get out of the internment camp seven times during his 22 months in Eire, but each time he was caught and hauled back. His eighth try, in September 1943, was successful.

3

First to Get Home Again

THE EAGLES WERE OFTEN out looking for targets across the Channel in the summer and fall of 1941. One such sortie, on October 20, ultimately produced a special inspiration.

The first of the Eagle Squadron fighter pilots to be shot down by the Germans and then to evade capture, make his way back to his old outfit, and get flying against the Luftwaffe again was Oscar Coen, a bright young former schoolteacher from Carbondale, Illinois.

"I was the first, but the fact that I got back all right and rejoined the squadron, after everyone thought I was gone for good, gave the other pilots hope that in similar circumstances they might be equally lucky," Coen recalled years later.

Fleeing from, hiding out from, and eluding the Luftwaffe, the Gestapo, and the SS for perilous weeks, Coen had time to contemplate the special circumstances that favored him.

> I was rather slight of build—above five and a half feet tall—and had a sort of boyish-looking face so that I appeared to be no older than 16 or so—at least two years under the conscription age. I soon realized that the younger I looked, the better chance I had of avoiding being challenged as I moved around under the noses of the Germans in France. I had soap and a miniature razor in my escape kit, and I would shave every morning, whether I needed it or not.
>
> I used to smoke quite a bit, so there were tell-tale nicotine stains on my fingers. I would scrub my hands often and thoroughly, to get rid of the markings. When I smoked I took to holding the cigarette between my lips the way the Frenchmen did, rather than in my fingers.
>
> I noticed that the Germans I would see on the streets paid almost no

attention to me. They apparently assumed I was a French lad going about my errands.

And the French, most of them, must have thought I was a young German. I couldn't speak either language, but under these conditions I didn't have to. Once I learned this fact I became much more confident. I discovered I could stand beside someone in a queue or in a room somewhere, and not worry about being too close to a potential enemy.

Oscar Coen joined 71 Squadron from Operational Training Unit 19, Sutton Bridge, April 19, 1941. He took part in the first combat operation by an Eagle squadron in July 1941, when Paddy Woodhouse, Bill Dunn, and Gus Daymond each shot down a Messerschmitt 109, and when Bill Hall bailed out into enemy hands.

On the morning of his own bail-out, Coen was flying wingman to Flight Lieutenant (soon to be 71 Squadron Leader) Chesley Peterson on a "rhubarb" search for suitable targets. "We had been flying over a cloud layer and started to let down through it to pick up targets," he said. "We were looking for ammunition trains on the way to Calais. Most of the freight carloads were supplying antiaircraft gun emplacements along the Channel. Those people were always shooting at us on our way back from our sorties. We were trying to cut down on their supplies a little bit."

Squadron Leader Peterson described the action this way: "Oscar was a real crackerjack, very keen and eager. We found a freight train near Lille, moving toward St. Omer. I told him, 'I'll take the engine, you take the cars.' I fired away, and the next thing I saw was a ball of flame. It looked as though an ammunition car had blown up and something had hit Oscar. I could see glycol coolant fumes streaming out from his plane.

"'I'm hit—I'm hit,' Oscar called out. 'I'm going down. I got to get out.' And then, his last words, 'I'm out.' I could see that his parachute was open."

Coen's account makes it clear that, starting with the first critical moments of chaos, his mind was focused:

> Pete and I pulled line abreast to have a shot at the train. I felt a thud and knew that an ammunition shipment had exploded and that debris had hit my plane. The radiator temperature started rising, and the cockpit filled up with glycol fumes and steam. After the rush of glycol fumes, the steam disappeared; my radiator was empty. The radiator temperature shot up at the speed of the second hand of a clock.
>
> The engine seized up just above the cloud layer, at 3,500 feet, and I bailed out. For bailing out over enemy territory, instructions were to delay opening the chute, if possible, in order to reduce the chance of being sighted and shot at from the ground. I knew that the clouds extended down to about 1,500 feet, so when I passed through the bottom layer it was time to deploy the parachute. I swung twice and hit the ground hard and sprained both ankles. Luckily I landed on plowed ground, and that softened the impact.
>
> I hid my Mae West and chute in rodent holes, pulled my trousers over my flying boots—they would give away my identity—tore off my wing

insignia and rank straps, and opened the front of my battle jacket so it
could look like an ordinary jacket. I rolled the jacket in mud a bit, too.
Then I took off, going away from the Channel. I knew that if the Germans
found the wreckage of my plane they would look for me between that spot
and the English Channel, so I headed straight inland. I never did see the
wreckage.

Coen came to an asphalt road and started following it, trying to put distance between himself and his smashed airplane as quickly as possible. He came to a railroad crossing where the gate had been closed to halt traffic. On the left-hand side of the road, outside of a guard shack, he saw an armed sentry. On the right side stood a building housing a small cafe.

I went into the pub. There were five or six people in the place. They
all stopped talking and looked at me. There was nobody behind the bar. I
rapped on it with my knuckles and the barmaid came out. I had plenty of
money on me. I pointed at a bottle and handed her a piece of currency. She
gave me change; I left the biggest piece for a tip and started sipping my
drink. By this time the other people in the room had lost all interest in me.

When I thought about this a little later, I realized that the key to the
whole situation was getting attention at the bar. That was when the people
quit being concerned about who or what I was, and just sort of forgot
about me.

Finishing his drink and leaving the place, Coen made a point of crossing to the left side of the road where the guard saw him coming.

He smiled and nodded and said something to me in German,
something not threatening. He did not even take his gun off his shoulder.
I just kept walking.

I reached the town of St. Omer almost before I knew it, and I used this
same technique again when I saw a German guard standing in front of a
tall stone building on the town square. I walked right on past. At the corner I made a left turn and looked back. The guard was leaning over,
watching me, kind of puzzled. I just walked on.

Coen's bruised ankles started giving him trouble, and his feet were quite sore. He was also becoming famished. Out in the countryside beyond the town again, he observed an entire family leave a farmhouse together.

I figured it must be some kind of a holiday, and Mom and Pop and the
kids were off walking to church. This was a chance to get something to eat
and some clothing.

I went inside and started rummaging around. On a wall was a picture
of British and French flags crossed, and to me that indicated that this was
the home of friendly French. Potatoes and meat were cooking on a stove.
I went into a bedroom to look for something to wear, and the back of my
neck started burning. A large man, the man of the house, was standing
there watching me.

I had a pack of Camel cigarettes, and I really startled him when I held them out. "Oh, Camall," he said. He studied me for 30 seconds, then motioned me to stay where I was. He spoke in French; I replied in English. Neither of us understood what the other was saying. He saw that I was hungry and gave me some potatoes and cheese. He excused himself for a few minutes and came back with another man who spoke English and was a local schoolteacher.

We talked for two or three hours. They asked whether I knew Franklin D. Roosevelt, and what did he think of the French? Did he consider them cowards for not fighting longer? I replied that I did not know President Roosevelt personally, but knew him to be a very accomplished man. I added that we Americans understood France's desire for peace.

They said they had friends who might help me. They bandaged up my ankles a bit, and noticed my pistol in my boot. Knowing that I would be unable to use the gun, I gave it to them. They made arrangements for me to go to a friend's house in a city. Meanwhile, they would hide me in a haystack for the night, and perhaps might take me to the friend's house in the morning.

They explained that this overnight stay in a haystack would break any trail that the Germans might try to follow. Early the next morning I was awakened by rustling noises of someone pawing about the haystack. Two men came and took me to a house in another town, Lumbres, 10 miles southwest of St. Omer. There, two women whose names I never learned—an older woman and her daughter-in-law whose husband had been killed in the war—told me I could stay for a couple of days. They had friends who could help me. They could not speak English, but the younger woman could write it, so she printed notes to me.

I had my own little room in their house, and I was never alone. The women took turns sitting at my bedside, and they were very attentive. When I awakened they would ask if I wanted water or something to eat, or maybe I had been having a nightmare? They also had me exercise in the backyard.

They gave me a Catholic holy medal to wear—a St. Christopher medal—although I am not a Catholic. More than 40 years later, I still wear it. A man came to take my picture, and soon they presented me with a new identity card with the name of Victor Thomas, showing that I was a Flemish worker. It even had an official seal. They burned my own ID card, and I was somewhat relieved. I had worried a little that if the Germans caught me and found my name was Coen they would assume I was Jewish. The world knew about the Nazi hate for Jews. I am not Jewish; I am of German and Irish descent.

It seemed a little odd that my friends gave me an English name, but it worked out all right. I was never picked up.

My friends learned that the Germans had found my parachute and Mae West imprinted with my name. They notified the Red Cross that I was now a prisoner of war. I was told the Germans sent in false information like this when they thought they could get away with it, in order to qualify for extra food and clothing parcels from the Red Cross.

Coen said his RAF escape kit included a miniature compass packed in the bottom of a toothpaste tube. Unfortunately, he added, moisture from the toothpaste put the compass out of commission. The way things worked out, however, he did not really need the device.

> Two welders who worked at a German air base drove me, in a truck enclosed in back, to Lillers, to a market on the town square. I stayed in a butcher shop where the owner and his eldest son, having nothing any longer to sell or buy, took jobs in a sugar factory.
> I slept upstairs, and stayed there perhaps a week. I had learned by this time that in German-occupied France you didn't go looking for the Underground. The Underground came and found you. I didn't know the names of most of the people who helped me. You did not ask for names or talk about them. This was essential. It meant that in case of capture by the Germans, literally you would be unable to give them information.
> Two British airmen joined me: Squadron Leader Harry Button and Flight Lt. George Barclay. We did exchange names and we talked a little, although it was dangerous to be together in one place. Finally the Underground near Lillers accumulated a group of us to be taken to the south of France, to Marseille.
> A tall skinny Englishman called Paul, supposedly a British agent, came along and sort of took charge. Without exactly knowing why, somehow we did not trust him. We never left him alone. Someone always stayed awake to keep an eye on him. We learned that the Maquis didn't trust him either. They suspected that he was a double agent for the Germans and Russians.
> We heard later that the Underground people had worked him over, had roughed him up badly. They beat him up for something. We never did learn for sure whether there was any real reason to mistrust him. Looking back, he helped us. He did us no wrong.

Paul's assistant in this dangerous game of "hide the fugitive but keep him moving" was Suzanne, an alert and attractive 19-year-old whose father was English and whose mother was French. Suzanne had remained in France when war broke out, to work with the Underground. She was clearly a brave patriot. None of her charges doubted her commitment to their cause.

The owner of the village butcher shop in which Coen, Button, and Barclay took shelter was another French citizen whose faith and devotion could not be questioned. Joseph Fardel died after the war, but the RAF men whose lives he had helped to save could not forget him. They gave a testimonial dinner in his memory in London in mid-1976. His widow Henriette and son Andrew were the guests of honor.

Joining Coen and his companions on the journey south were RAF Spitfire pilot Alex Nitelet, a Belgian, and two British soldiers who had been in hiding in France since the Dunkirk landings. When it came time for them to cross a river into the sensitive restricted area around Abbeville, they had to use a special pass, and they were instructed to cross the bridge singly. Paul and Suzanne left them temporarily.

"Our Free French friends handed me the ID card of a lay priest, whose name I can no longer remember, to show at the check point into the new zone," Coen said. "I wore slacks and jacket and low shoes, but had no clerical collar. When I showed my card to the German guard he looked at it, smiled and said '*danke schön.*' I had no trouble at all at the crossing, and neither did the others. My identity card was returned to the priest. I was Victor Thomas once again."

In the nearest town they all went separately to a theater where a James Stewart movie was playing. When Coen walked out after the show, Suzanne was across the street waiting for him.

From this point on, the transportation was varied. In a horse-and-buggy interval, Coen made himself comfortable in a bundle of hay. Suzanne and Paul managed somehow to keep an eye out for everyone.

> Generally we rode trains. In getting ready to buy a ticket, we would practice saying the name of the town we were going to. Aboard trains, we always tried to sit beside someone of our group.
>
> One time when we were moving as a group we had to walk, at a fast pace. We covered 26 miles from sundown to sunup. We caught a train early the next morning and were on our way to Paris.
>
> There three of us were placed in an artist's apartment. There were racks down one side of a room, covered with drawings and finished and partially finished canvasses. We would take some of our meals at a friendly Parisian restaurant, where they knew who we were. After two or three days we went by train to Marseille, where the group was split up and we parted company with Suzanne and Paul. Suzanne was small and delicate and very pretty, the last person one would suspect of being involved with the Maquis. She was completely unflappable. If she had not been in the party I don't think I would have gone very far with Paul. And if we had been captured, I would have gone to a prisoner of war camp—but she would have been shot.
>
> I was sent to Nîmes, northwest of Marseille, and was told that the others were coming, too. We stayed at what was a smugglers' base between France and Spain. Outside it looked like an ordinary warehouse, but inside it was quite an apartment—with a housekeeper who did the cooking.
>
> After we had been there about a week we were told to be ready to leave in the morning. We were given a loaf of bread and a bottle of brandy and were advised to "save the brandy until it gets cold—you will need it heading into the Pyrenees."
>
> A smuggler—we did not know his name—was our guide crossing over the mountains into Spain. We started about noon, following a footpath and walking fairly steadily. It was raining and cold, and the ground was frozen. We arrived in a village about 3 A.M., but dared not attract attention by going to the rail station. We were told to lie down on the ground and keep quiet. A train would be along in about three hours. The rain was freezing. I have never been so cold in my life.
>
> A short train ride brought us into the city of Figueras, Spain. A woman who met the train introduced herself as the wife of the British Consul and took me home with her. The others in our group I did not see again until we all met in the British Embassy in Madrid.

> My Figueras benefactress led me to her back porch, brought me a robe, and told me to take all my clothes off. "I don't want them in the house," she explained. She probably thought, with good reason, that they were lousy. They were no loss. After a good bath, and dressed in clean clothing, I felt like a new man.
> I stayed in Figueras for two days. Each morning my hostess and her chauffeur would drive out to try and find groceries for them and her husband. Food was in very short supply, difficult to buy.
> Next was the train ride to Madrid, and in the British Embassy there for the first time I felt safe. All of the fellows arrived within a day or two. We lived in a Nissen hut in the Embassy grounds, and we did our own cooking. Also, we could draw money on our RAF pay. The Embassy people told us, "You will stay here until we get you out. We will work it out with the Spanish. There is no point in getting into trouble trying to do this on your own."
> Arrangements were made for us to go to Gibraltar by train with a Spanish guide. The guide kept trying to tell me something all day, but all I could pick up from his attempts at English was "Pearl Harbor." Not until we crossed into Gibraltar this December 8 did we learn from the border guard what our Spanish guide had been trying to tell us: the Japanese had bombed Pearl Harbor.

In Gibraltar Coen was surprised and delighted to encounter three former colleagues of 71 Eagle Squadron. John Campbell, Don Geffene and Art Donahue, bored with the routine Channel patrol missions that were their assignment in England in the late autumn of 1941, had volunteered for duty in the Middle East and now were awaiting transportation to Malta as members of the RAF's 258 New Zealand Squadron. They welcomed Coen as a man returned from the dead.

A chronic shortage of air transport at Gibraltar irked the newly escaped-from-France fighter pilots, whose travel authority was on a space-available basis. The restless airmen began seeking their own ways home. Coen stumbled across some Australians of an embarkation unit he had met aboard ship on his first trip to England, and found that they now were members of a Sunderland bomber crew en route to Britain. He told them of being stranded here. The sympathetic airplane commander responded, "I can't suggest to anyone that I have room for a passenger. That would put you on the lowest priority. But as a member of our crew I can get you to England on Christmas night."

On arriving at the military manning pool outside London, Coen went through the debriefing procedure as though he were a regular crewmember.

> In the replacement pool my batman got my clothes for me. I went to the Group Captain to inquire about orders, but he was having tea and couldn't see me. It was a beautiful day, the weather was good, I had my clothes and they were pressed, and I had money from Spain. So I got on the train and went to London, and then back to 71 Squadron. Pete asked me if I was ready to fly. I was, so I went back to work.

Six weeks later I got word that the MPs were looking for Coen because he was AWOL—absent without official leave—from the manning pool. Robbie, our intelligence officer who later became Lord Martonmere and was Governor of Bermuda, got on the phone and straightened it all out.

Oscar Coen went back to work, all right. He fought bravely in numerous air battles, became an ace with 5½ enemy aircraft destroyed, and ultimately retired from the U.S. Air Force as a full colonel.

4

Land of Dear Souls

THE EAGLES UNIVERSALLY HOLD fond memories of their welcome during the first years of conflict. Some, like Gene Tobin, spent free days and evenings in the glamorous company of war correspondents, movie people, and other notables. Others enjoyed the cheer and hospitality of ordinary people who felt and showed extraordinary gratitude for the staunch presence of their American "mates." Many got to see Britain at what many Britons and Anglophiles regard as the blessed plot at its best—the lush, serene countryside and an endearing, traditionally gracious way of life.

Beset by shortages, personal losses, and usually mindless bombings, rural Britain retained seductive charms of landscape and custom that poets, painters, and storytellers have celebrated and Britons and visitors have loved. As the war progressed, Britain did become, in some eyes, "the world's largest aircraft carrier," and in others, a land of khaki forests and metallic meadows. The landscape was dotted with airfields, depots, tent cities, and areas whose appearance hinted at preparations for unusual means of warfare. That notwithstanding, the country had more than enough true forests and meadows, tranquil, winding rivers and quaint, snaking roads to recall a Britain at peace. To leave tense, hectic, gritty London for the country was to shift the gears of existence to a more gentle pace and, very often, to a more genteel style. Here also was, perhaps, the true-bred Briton's element, the culture out of which he could nourish his individualism, his independence and, most important, his vital eccentricities.

Vernon A. Parker was called "Shine" by his friends, including the English whom he came to know especially well and whom we will meet through his memories. Curious, attractive, affable, and adaptable, Parker easily settled into a variety of milieus and viewed the people he met with a fine perceptive eye. He was also moved to feel a sense of obligation to pay them more than the superficial tribute of the passing hour.

At the time that he made his commitment to join the war, well before Pearl Harbor, Shine Parker was a civilian pilot at Del Rio, Texas. Red tape discouraged his attempts to join the RCAF directly in Canada, but as often happened, the savvy operatives of the Clayton Knight Committee greased the wheels to secure a talented pilot for the Royal Air Force. Their tactic was for Parker to sign up with British Commission Ferrying, Ltd., at Love Field. The BCFL was, of course, a training school for the RAF.

Parker arrived in England in June 1941, under conditions that were hardly hospitable. The transatlantic voyage had been planned for 17 days—it took 31. Food aboard ship was scarce, and the water allotment for all purposes was one quart per man per day. Rations in England, once he arrived there, were slightly better but, he wrote, "I was still hungry when I transferred over to the Americans, and then had my first good meal in almost a year and a half."

What he lacked in the way of food he seems to have made up for in companionship—with some good food thrown in. He had the good sense to take advantage of a special program. Here is Shine Parker's account of found treasures:

> A home away from home—this was the slogan used by a group of ladies in London headed by Lady Frances Ryder. They were closely associated with the American Eagle Club, as well as with other clubs for foreign officers stationed in the British Isles and serving with the Royal Air Force and other branches of the service.
>
> When an officer had leave in excess of forty-eight hours, with no place to go, he could contact these ladies for an interview. They were adept at sizing up each individual's character and personality. Each man was asked which particular part of the British Isles he wished to visit and whether he preferred city life or a sojourn in the country—the Midlands or the coastal regions. He was questioned also about his hobbies, such as golf, fishing, riding or hunting.
>
> The replies were analyzed carefully so that the American could be matched with a family—usually one that was well-to-do—capable of entertaining him for the extent of his leave. All that was required of the officer was that he bring his ration card and, if indicated, a bottle of the best booze for his host, as well as his ration of chocolates for the younger members of the family.
>
> I stayed with a number of families, as arranged through these ladies, and often visited a particular family in Scotland. I became virtually an adopted member of the household, mainly because I was about the same age and personality as a son they had only recently lost in the war. Every family was one of dear, warm, charming, and generous people.
>
> On my first leave, I requested a place in the country in Scotland, since I had been all over England and Wales. The ladies and I discussed several families in or near Edinburgh and Glasgow and studied pictures of their

homes. I chose the family of Colonel W. E. Peel, at Haddington, near Edinburgh.

I made the overnight trip between London and Edinburgh on the *Night Scot*, a very comfortable train, which before the war and its blackouts, I was told, averaged 110 miles an hour. A large Humber automobile and its driver met me at Waverly Station in Edinburgh and took me the nineteen miles to Haddington and the Peels' estate, Eaglescairne. As we drove up the circular driveway and approached the huge two-story mansion, a portly gentleman came bounding down the stairs.

"Good morning, Colonel Peel,' I said, 'I'm *Left*-tenant Parker." I extended my hand.

It was refused.

"I am Barnes, the butler," the slightly ruffled and embarrassed man said. "The Colonel has not yet arisen."

Barnes ushered me into the drawing room and said breakfast would be served at 8 o'clock. Soon the house came alive. The first in to meet me was Doriel, about 19, bubbly, exceedingly cheerful, and overwhelmed at meeting her first American, especially a Texan. Then in came Arimente, about 24, a beautiful blonde, who was much more reserved but just as interested in hearing about Texas cowboys and ranchers.

And then Colonel Peel, a fine, strong man, very distinguished looking, slightly greying at the temples, with a military bearing, tough, slightly stooped. He limped from an old wound in his left leg. He had a clear, booming voice but a gentle manner and did a lot of wheezing and muttering. I could tell immediately that we would become very good friends.

We were at breakfast when Mrs. Peel came in. The others merely bounced up an inch or two as she was seated; I stood up. She gave me a long once-over from head to foot, looking so stern I wondered momentarily if I should get out and go back to London. Then she smiled ever so warmly and asked me softly if I was a cowboy. The ice was broken and I was accepted. Before breakfast was over, I didn't care if I never returned to London.

The rest of the week was spent relaxing, getting acquainted, and visiting the countryside, including Robert Burns' homesite just a few miles away—not his birthplace at Ayr, but where he lived and worked in later life.

We talked mostly about their son, an RAF pilot killed in action just two or three months before my arrival. They showed me his pictures, and there was the resemblance of brothers between us. It became increasingly clear as time passed that I was being accepted as if he had returned home on leave. Several times I was absentmindedly called by his name. On my first visit I was given a guest room, but for every visit thereafter I had his room. I was the only person allowed to sleep there.

There were two other children. The oldest daughter, Argenta, was a barrister living in London. She was also the first woman Member of Parliament. John, the next oldest, was a captain in the Royal Corps of Engineers and was stationed in the Orkney Islands, north of Scotland. Later, when it was known that I was coming, John would try to get leave to be home when I was there. If Argenta could not come, I would meet her for dinner when I returned to London, to give her the latest news of the family.

John and his wife, Joan, came to Eaglescairne on my second visit. Joan was a beautiful, trim, athletic type of girl with long, flowing, golden hair. She had a way of kissing me on every occasion, but always and only in full view of the family—never once when we were alone. This never caused a raised eyebrow, not even from John.

Joan became pregnant on that visit. The British are frank in discussions on any subject, and so, on frequent occasions, the conversation—from all except Barnes—was along the lines that "we'll never know if that is John's baby or Shine's." Joan happily joined in, neither confirming nor denying. Everyone thoroughly enjoyed the topic except me. I was embarrassed and a little scared, even though there never was an occasion when I could even remotely have done anything to be considered the culprit.

Landowners and farmers were required to register livestock and farm products. Colonel Peel raised chickens and pheasants, but he refused to register them. As was typical of most Scotsmen, he resented any restrictions imposed by Englishmen. Because of his refusal to register, he was unable to sell chickens or eggs, so he gave them away.

Twice a week we drove into Haddington and Dalkeith for supplies. The Colonel took his supply of eggs in a small crate. As we stopped at the greengrocer's, he put six eggs into his baggy coat pockets—one for each member of the family—no more. The barber got two eggs, the butcher four, the chemist [druggist] three. It was amusing to see this hulking man with his pocket full of eggs, walking down the street, wheezing and muttering to himself, passing out eggs. Why not, when the two of us got more than our share of brandy or sherry!

Golf was the Colonel's major outside interest. He said it was for exercise, but I wondered about this after a few trips with him. Usually we drove to the old Longniddrey Golf Club and Golf Museum three or four times a week. It was a wonderful old, dark, stuffy place where the whole history of golf unfolded before your eyes. In those dusty old glass cases were balls and clubs with the dates and names and scores of those who had used them. Walter Hagen [American winner of the 1922, 1924, 1928 and 1929 British Opens plus other championships] had a special display. Some of the very old balls were as large as basketballs, and some even had feathers attached. Some of the very early clubs looked like tree limbs with a crook in them.

I suspect that the Colonel's interest in the game stemmed partly from the fact that there was no rationing at the bar. Members and guests were served all they cared to drink. Over the period of months that I was going there I never did complete nine holes. We would tee off from No. 1, go directly to No. 3 and from there cross to the ninth hole at the clubhouse. After a quick brandy, it was off again to 1 to 3 to 9, in a pattern that never changed.

Several times at Eaglescairne, there were other guests of various nationalities. A very young Free French officer was a weird one who was never asked back. He thought he could mesmerize animals and make them do his bidding. He worked on the horses, cows, chickens, a couple of ducks, and a pheasant until he nearly drove them crazy. His visit ended when he tried his gift on Barnes. The butler kept calm for about an hour, trying to go about his work and to ignore the Frenchman. But the mesmerist kept pestering him.

Finally Barnes admitted, "I am under your control—but let us please get out of the house."

There followed a great commotion from the wellhouse, a lot of banging around. Barnes returned to the house alone, unruffled, and prepared tea as graciously as ever. Louis came back about an hour later with a lightly swollen eye and a cut lip. He packed hurriedly and left for Edinburgh. Barnes smiled and winked at me as the Frenchman drove away. After that, Barnes and I were quite friendly, and I usually had an extra ration of sherry by my bedside at night.

Another interesting guest was a small, dumpy Belgian general who could speak English, as he put it, "very few." Even so, he had no trouble making his intentions known. I was expected to stand when he entered the room and not to speak to him until he had spoken to me. Usually, wherever I was seated he wanted to sit, and whatever I was reading he wanted to read—and *now*! Barnes, the keeper of the keys to the liquor cabinet, cut off the general's supply of refreshments and at the same time increased mine. The old gent learned that for a good glass of brandy he could be nice to me.

After about 14 months, the war hung heavy over our heads. I had already given up any thought of seeing the States again and lived from day to day. Every mission, every day I expected to be my last and was surprised to find myself accepting this calmly. We had lost many friends in combat, and the invasion of England appeared to be a certainty, just a matter of time.

I learned I was to be posted from 121 Squadron to an instructors' school as a rest from combat duties. At first I rebelled at the thought of leaving my American friends; then I accepted the idea of relief from the dreary routine.

My new post, No. 32 IFTS—Instructors' Flight Training School—was at a small civilian field near Worcester. It had no accommodations. The housing officer gave me a list of homes near the base and in Worcester that I could visit and from which I could make a choice. I wanted a really quiet family residence and thus selected The Old Vicarage, an ancient two-story, red brick building with stained glass windows and a steep-slooping roof, completely enclosed in a ten-foot-tall red brick wall.

No one answered the bell at the front door. I walked around and found a double gate at the rear, partially opened. In the yard inside, a small automobile stood jacked up, with parts lying all about and two feet extending out from under the car.

"Is the vicar home?" I asked.

The feet thrashed about, and the figure of a woman 35 or 40 years emerged. "Hell," she said, "no vicar has lived here in years. You must be the damn bloody American that called."

Thus began the wildest, most hilarious two months of my life. I was trapped. My hopes for peace and quiet were quickly forgotten. We went into the house, opened the bar that never closed and drank and chatted until Alex Peters, the husband, arrived. I never learned what he did; some sort of secret government work, I suspected, but he never discussed it. He seemed to have plenty of money and expected to party every minute he was off duty.

After a light dinner and a few more drinks, Mrs. Peters—Eileen—called a friend, Rhea Windom, in nearby Malvern, and asked her to come to the King's Arms, a quiet little country pub between Worcester and Malvern, to meet this *dahling* American pilot. Rhea brought the whole Windom family and called a few other nearby farm families to greet the flier from Texas, "over here just to help us win the war."

It was a wild night. When I awakened at noon the next day, I found that Rhea had also moved into the Vicarage, where she stayed until I left two months later. Separate rooms—yes. Rhea was classified as a farm worker on her father's farm so that she could be exempt from the draft.

The Windoms raised apples and also hops on contract with Bass Ale. A nearby Italian prison camp provided more than enough labor for the farm. About sixty Italian prisoners of war were transported daily to the farm under the guard of one old man with a shotgun. The gun was never loaded. The Italians knew this, and it worried me until one of them said to me, "Who cares? We have no place to go. Who would want to go to Italy?"

Rhea almost lost her exempt status when she moved to the Vicarage. It was necessary for her to live on the farm at least on weekends. The RAF base shut down from Friday afternoon until Monday, so I spent the weekends on the farm with Rhea.

Upon graduating as a flying instructor, I was posted to No. 16 EFTS—Elementary Flight Training School—at Burnaston, again a civilian field pressed into military use, with no quarters. Again I studied a list of available homes, this time definitely with peace and quiet in mind. My choice was a modest, exceptionally clean and quaint home—the third in a row of eight houses—so typical of Derby. The owner was Mrs. E. E. Peck, a very sweet, kind, white-haired lady who had retired after almost forty years of teaching school. She had now been called back to teaching half a day, three days a week because so many of the younger teachers had been drafted.

As dinner was being prepared that first night, I received my briefing on how things were to be. First, there would be no smoking in the house and no drinking except for a small glass of sherry before bedtime. No girls could visit me in the house or in my room, except that on Sunday afternoons I might have a female guest for one hour, at tea time.

This sounded dull and straight-laced until I saw my flying schedule gave me little free time. Flying would keep me away three nights a week, and one week each month I would spend at Abbots Bromley, an auxiliary field for Burnaston.

EFTS provided army enlisted personnel with a minimum of sixty hours of basic flight training and four hours of solo flying, as a preliminary to glider training school.

The RAF and the British Army got along with each other in about the same manner as did their American counterparts. We were expected to wash out 50 percent of the trainees. This meant that with each new group, for the first few days we were to do all the aerobatics in the books. Any trainee taken ill on a ride was immediately eliminated. We flew de Havilland Tiger Moths and Miles Magisters. The Tiger was a beautiful little plane for aerobatics, even at very low altitude—very stable and reliable. I enjoyed this part of my flying career most of all.

At breakfast in the Peck home I would find an orange by my plate. For most British this was a rare treat because, when a few oranges did come in from South Africa, they usually were given to children under ten or to persons over sixty-five. When Mrs. Peck was at the greengrocer's, he would slip her an extra one "for the American." She refused to eat her own because, she said, as a pilot I needed it more than she did.

Each morning, before setting out for the base on my bicycle, I had to stand inspection, like a small schoolboy. Unless my brass buttons were shining, my shoes were spotless, my uniform was neatly pressed and clean, and my hair was properly combed, I was sent back to my room to prepare for a second inspection.

Sir Michael Assheton Duff-Smith, a Welsh baron, had been assigned as one of the original members of 121 Eagle Squadron as the intelligence officer. It was debatable as to whether "Mike," as we called him, had transformed us into a British unit or we had changed him into an out-and-out Yankee. It didn't matter because Mike had become one of us. Everyone loved, admired, and respected him. He, in turn, was fond of his American friends.

William L. C. "Casey" Jones, of Baltimore, was my roommate. At least once a week, Mike, Casey, and another British intelligence officer, Jenkins, and I went to town in Mike's little car for a bit of pubbing and dinner and perhaps an American movie, preferably a Western. One night, Mike told us that his good friend, Adele Astaire (the sister and former dancing partner of Fred Astaire), now Lady Cavendish, had asked him to send a few Americans over to her place—Lismore Castle, in Eire—for a rest-type leave.

There would be a few problems. Since Ireland was neutral, we must wear civilian clothes. We should have passports. Prime Minister Famon de Valera had closed Ireland's northern border to Americans because troops coming into Belfast had gone south and created quite a stir. Eire was, of course, unsympathetic to the British war effort, but traffic between England and Eire never was stopped. The ferry between Holyhead, Wales, and Dun Laoghaire, near Dublin, regularly ran twice a week.

In London, we were denied passports because we were not British subjects. Instead, our RAF identity cards were stamped to serve as passports. To our surprise, we had no trouble with Customs. Our RAF IDs were stamped as though it were normal procedure. When our British accents slipped so that the Irish discovered we were Americans, we had a ball. Every Irishman we met had been to America, had relatives in America, or was going there as soon as possible.

Wandering around Dublin, we began to realize that we were no longer in a country at war. Except for matches, gasoline, and Scotch, there was no rationing. Cages displayed big hams, roasts, and fish in their windows. After more than a year in England being limited to one egg a month, fish sometimes once a week, and brussels sprouts and carrots every day, we found all this hard to believe. The main entree everywhere was steak and eggs. When we asked a waiter if we might have two eggs, the answer was "Sure and why not?"

We had heard of the fuel shortage in Ireland, and it was verified for us on the train trip from Dublin to Mallow, in southwestern Eire. We stopped no less than ten times so that the passengers could scrounge for cow manure, sticks, or anything near the tracks that would burn. The farther from Dublin, the scarcer the coal.

At Lismore, we were greeted by Adele as though we were old friends. Within an hour, we felt that we had known her always. She was the most gracious hostess I had ever known—a doll, and pleased to see a few Americans again. Having danced most of her life, at 42 she still had the grace, poise, and figure of a sixteen year old.

Lord Cavendish was a semi-invalid, and we did not meet him until dinner. Adele told us so much about him—about his vitality and his broad interests—that we knew he was like her, in every way full of life. From the start, he had us on a first-name basis: He was Charlie, we were Shine and Casey.

Adele described the old castle as "our 200-room home with one bath." Actually, seventeen rooms in the east wing had been modernized as living quarters. The old banquet hall and other rooms remained intact, in their original condition. We spent many hours exploring the fantastic structure, even to its dungeons, in one of which was a pile of bones several hundred years old and still in chains.

In his younger days, Charlie had bred cob horses for racing. His best friend was Paddy Walsh, a mortician at Cappoguin, eleven miles away. Paddy was a typical happy-go-lucky Irishman, even to his bald head and bold grin. Like Charlie, he loved horses, and the two of them would go to races in Paddy's hearse, for which petrol was not rationed. Paddy would load Charlie in the back and they would drive all over the countryside, with no one ever questioning them.

Charlie's affliction was a jake leg, a paralysis that is caused by liquor, so he could not touch a drop. Paddy drank poteen, the Irish potato whiskey, in a quantity sufficient for both of them. Every other drink, he raised his glass and said, "This one is for you, Charlie boy."

Wednesday was marketing day, and Adele invited us to come along. We walked through the lovely green countryside—the weather was very much warmer than in England—to a quaint little store where we were ushered into a back storeroom. The aroma was out of this world. Slabs of cured bacon and hams and sausages were hanging from the ceiling. Adele gave the man her list, and we sat on sacks of sugar and beans and had drinks on the house.

On Thursday night, Adele had a dinner party so we could meet a dozen of her friends. We had fresh salmon steaks from their own traps on the Blackwater River, a source of a large part of their income. Among the guests were Mrs. Jameson, of the family that produced Jameson's Irish Whiskey, and the novelist, Daphne du Maurier. Mrs. Jameson drank her whiskey neat and admonished us against putting anything into it. "If it needed anything," she said,"we would have put it in when we bottled it."

We spent as many as three or four hours a day in Charlie's bedroom listening to his tales of cob racing, salmon fishing, and old Lismore Castle. He tried hard to persuade us to get him a drink, but Adele kept the key to the bar and guarded it every minute. One drink could put Charlie off his feet for a week or more.

He told us that when he was 21, he worked in New York for J. P. Morgan and Company. He lived in an exclusive hotel and had a beautiful blonde neighbor whom he finally managed to meet. He took her out to dinner and then returned to the hotel, looking forward to a night in her company.

"But her boyfriend was waiting in her room," Charlie recalled. "He turned out to be Legs Diamond, the gangster. Two bodyguards were with him.

"We had a short talk, and that same night I moved to a new hotel across town."

We had been granted seven days' leave. Without our knowledge, Adele wired London and arranged for a seven-day extension "because of transportation problems." We were delighted, of course. It was a fourteen-day vacation such as Casey and I could never duplicate. Although we agreed to come back in six months, we were most reluctant to head back toward dreary England and the war, blackouts, and rationing.

A month after our return, Casey was a prisoner of war. Six months later, Charlie was dead, and the Air Ministry refused any more leaves to Ireland.

Adele came to London once while I was there, and I took her and her friend—a daughter of Marshall Field, of Chicago, married to a British admiral on duty in the Far East—to Paul Xenia's Russian Club for a steak dinner. Paul had been a commander of No. 99 Bomber Squadron, which was made up of American volunteer pilots, in World War I, and he was especially partial to Eagle Squadron members. These were the only steaks I ever bought in London.

In 1975, from his home in Floydada, Texas, Shine Parker spoke not only for himself but for many other Eagles and American warriors who were drawn into close friendships with their British allies in the first years of combat. "By offering the Yanks a special welcome," he said, "they contributed much to the war effort. Sometimes a hardship was placed on them, but they were ever helpful and never complained. They should be recognized in some way."

In England, Scotland, Wales, and Ulster, and even neutral Ireland, the young Eagles found cultures both similar to their own and different enough to require bridging a gap. On both sides, these meetings of peoples called for an extension of understanding and receptivity. Far more often than not, the bridge was created and was strong. Americans and Britons alike might first have found confirmations of stereotypes, but whatever national clichés seemed to be confirmed at first impression, deeper personal harmonies soon developed. Instead of a clash of cultures, there resulted a mixture of feelings, including loyalties, between the outgoing, often lonesome American youths and the friendly, if unconventional, Brits that was rich and warm.

Many Americans who were shot down over Nazi-occupied Europe would learn from the diverse peoples there who helped them how much a common cause can stimulate brotherhood and uncommon courage. Yet even the first of those Americans who were forced to become strangers in strange, dangerous lands had already grasped something of that process within the shores of Shakespeare's "land of such dear souls."

5

Battles against the Japanese

ALTHOUGH 1941 REPRESENTED a time of trauma for the downed Eagles, in the late autumn, and on into the winter, the air war over cold, rainy, wind-swept England and Northern France slowed almost to a standstill. Patrols became progressively routine, and a restlessness set in among some Eagles.

Oscar Coen's welcoming party in Gibraltar—Californians John "Red" Campbell and Don Geffene and Minnesotan Art Donahue—had sailed from England November 3, 1941, which happened also to be Campbell's twentieth birthday. They had orders to 258 Squadron and a vague understanding of their "overseas base." "At the time we knew our destination only as Egypt, which was a nice warm place," Campbell said. However, events were soon to dictate a radical change in compass headings for these restless Eagles. As Campbell recounted:

> We ended up as a volunteer mission that was to go to Gibraltar and board the carrier *Ark Royal*, from which we would fly off, stop at Malta, and then by steps proceed up through the Middle East to the Russian Eastern Front. There we were to replace the RAF Wing attached to the Russians as a token force.
> We were equipped with cannon-firing Hurricane IICs. We had been loaded, along with Hurricane squadrons 242 and 605, aboard the H.M.S. *Athene*, a seaplane tender. The airplane wings, of course, were dismantled.
> At Gibraltar, half of the Wing—242 squadron and half of 605—were loaded on two carriers, taken into the Mediterranean and flown off to Malta, where they stayed. On the way back to Gibraltar the *Ark Royal* was torpedoed and sunk, so we sat on Gibraltar for seven weeks. We were there when the Japanese attacked Pearl Harbor December 7.

During the long stay at Gibraltar, Campbell and Donahue were much amused with the activities of their teammate Geffene. "Don was a sophisticated, attractive man with a magnetic personality," Campbell said. "He was one of the few Eagles who had married before getting into the RAF. He had been married and divorced. When he walked into a room, you could see all the women in the room start watching him."

> In Gibraltar, the only females were a few Spanish women, who came across the border in the daytime, worked in the shops, and then went home. Geffene was the only one of us who could find feminine companionship there.
>
> He set himself up with a girl who worked in a butcher shop. A young navy lieutenant from the *Athene* found out about this and asked Geffene how he managed it.
>
> "Just put your money on the table," Don told him. The lieutenant tried this, and got chased down the street by a very angry Spaniard with a butcher knife.
>
> When we first came to Gibraltar, there were half a dozen British WRENS there. We called their mess the Wrennery. One of the squadron commanders said that if we would promise not to insult the ladies he would arrange a party for us.
>
> Geffene learned that the CO planned to fix himself and another pilot up with WREN dates. Don promptly went down to visit the British battle cruiser in the harbor and told the officers that everyone there was invited to a big party at the Wrennery. The squadron leader arrived at the party he had arranged to find the place filled with Navy types.

Every day during their long stay on Gibraltar, pilots flew their long-range Hurricanes on low-level patrols along the Spanish coast, searching for the Luftwaffe's Focke-Wulf Condors, based at Cadiz, which had been menacing shipping on the Atlantic. On one such mission Geffene reported engine failure and made a forced landing in Spanish Tangiers. Although this supposedly was neutral territory, Campbell raced in and destroyed Geffene's Hurricane on the ground with cannon fire. "I wasn't about to let the Spaniards bring their German friends in to inspect Don's plane," Campbell explained. There was talk of international repercussions. Campbell was not allowed to fly any more missions from Gibraltar.

Geffene escaped injury but was interned by the Spaniards. Security was lax and he soon managed to escape. Friends said the Governor General's daughter, charmed by the American pilot, arranged for a visiting war correspondent to smuggle a gun and maps to him. Making his way somehow to Colombo, Ceylon, Geffene soon was flying for the RAF again. He was killed in action during a Japanese raid on Colombo April 5, 1942.

Campbell and Donahue moved on with 258 Squadron, on the seaplane tender *Athene*, to Takoradi in Gold Coast, the British African possession lying between the Ivory Coast and Nigeria. With Hurricane wings refitted to the aircraft, the squadron flew eastward across Africa to Khartoum, Anglo-Egyptian Sudan. Here the squadron was relieved of its Hurricane IIC equipment because the cannon-firing aircraft were needed urgently in the Northern Desert. The pilots of 258 Squadron took possession of machine-gun-equipped IIB Hurricanes, instead. The British aircraft carrier *Indomitable* ferried the squadron from Port Sudan to within 300 miles of Java. The pilots flew off to Batavia for refueling, and then to Sumatra for more petrol. On January 29, 1942, they reached their immediate destination, Singapore.

Japanese forces, following up the early-December bombings of Hawaii, Singapore, Hong Kong, Guam, Midway, Wake Island, and the Philippines, had landed on Siam (Thailand), Malaya, and North Borneo, and on Luzon and the Lingayen Gulf in the Philippines, and then in quick succession had captured Wake and Hong Kong. Pushing on strongly in January, they occupied Manila, invaded Kuala Lumpur, and made landings in Burma, Balikpapan in Borneo, and New Britain and New Ireland in the Bismarck Archipelago east of New Guinea.

The Hurricanes bolstered the small, desperately beset air defenses of Singapore just as the British forces were withdrawing into the beleaguered city and demolishing the causeway to the mainland. "All hell had broken loose," Campbell said. "For us it was a case of too little, too late. No ground support, no proper maintenance, but we did put up a good fight.

"Having no maintenance, we did our own. We lost some planes, banged some up on landings. The largest squadron formation we were ever able to put up was seven, in our first major combat February 1, 1942.

"The Japanese had two squadrons of 27 bombers—54 big planes—and three squadrons of fighters. It was the first time I ever had attacked a bomber formation as such; before that the bombers had always been singles. We lost three planes that day. Art Donahue shot down his first Zero."

Campbell provided this account of that engagement:

> The Singapore warning system was bad. We would take off heading south away from the Malay Peninsula to get altitude. Tallyho on the bombers came just north of Singapore, and our squadron—all seven of us—dived into them.
>
> The lead bomber in the right-hand section attracted my attention. My first burst hit in the upper rear gunner's position, putting him out of action. I could see him lying back over the gun, which was pointing straight up.
>
> I continued my attack all the way in until I was dead astern. From this position I took several short bursts with no effect. I then realized that since I was holding my gun-sight square on his tail cone, my guns were about six feet below and about eight feet out from my line of sight, so most of my bullets were passing below him.
>
> I pulled up to correct this, and his two wingmen, whom I had been ignoring, opened fire on me. They had not been able to depress their guns sufficiently to do this before, without hitting their own tail surfaces. I got the hell out of there and then pulled up and made a quick belly shot.
>
> By this time there seemed to be fighters and tracer bullets all over the place. As I dove away I saw one bomber drop out of formation and blow up. I could not tell if it was mine or the one Denny Sharp destroyed.

In the next moment I noticed two Zeros chasing a Hurricane. They had him boxed in, and I started to dive to help. The Hurricane pilot, probably Bruce McAlister, who was lost that day, turned as if he was going to attack me. In the confusion he must have thought I was an enemy aircraft. I turned away so he could see my lower surfaces, where the markings were easier to read, and he turned back.

The two Zeros closed in on him and he dove away, appearing to be out of control and trailing white smoke. I continued my dive and observed hits on the Zero on the left side, who appeared to be the section leader. I continued firing into about 50 yards when he pulled up vertically, engulfed in flames. I had to do a very violent snap roll to avoid hitting him. I was doing about four hundred miles an hour when I snapped it. Something must have bent; that Hurricane never did fly straight again.

When my eyes came back into focus, I noticed that the other Zero was flying alongside of me, parallel, and we were just looking at each other. I turned into him, and for some reason he turned right in front of me. I opened fire and saw some hits. I fired only a few shots and suddenly had a feeling that there was something behind me. By the time I had cleared my tail, he was long gone.

I dove for the deck to make myself a more difficult target. On the way back I used the little ammunition I had left on some troops along the road. For this engagement I claimed one destroyed, two damaged.

We flew operations out of Singapore for about a week. Then the Japanese started to shell the airfields and we moved back to Sumatra, to a former KLM Airlines civilian airport 10 miles north of Palembang. The Japanese landed on Singapore February 7.

I remember a particular encounter over Palembang, when the Japanese were coming in to raid our field. We scrambled, and the meeting broke up into a general dogfight. Three of them came after me.

The Hurricane, climbing full power at 220 miles an hour, would simply leave a Zero behind. The Zero could never catch up. This time we did not have altitude and were just off the deck.

I started to lose one of the Zeros in my rear-view mirror, and began weaving to see where he was. All three were closing on me, and as one opened fire I did a 180-degree turn— during which he scored several hits on me—and fired head on.

As I came on top I could see one-fourth of his engine cowling ripping away. I went over him and back up in the clouds and had no idea where he was. Luckily for me he crashed right off the airfield—rolled and went right into the deck. The ground crew confirmed that it was a definite, not just a "probably destroyed."

Although he was engaged in combat operations for less than seven months, Campbell was credited with shooting down at least six enemy planes. He was respected among his squadron mates for his cool headedness and remarkable steadiness under fire. "Bravest guy I've ever known," one pilot commented years later. Another, Terence (cq) Kelly, recalled in his book, *Hurricane Over the Jungle*, that on February 3, 1942, "we were expecting a flight of Hurricane replacements from Singapore and they turned out to be Zeros, strafing us. I began to run and then I had to stop because a few yards away I saw something which to me seemed quite incredible. Red Campbell was standing calmly, revolver in hand, aiming at the next navy Zero. It was useless, of course, but not a gesture. One does read of men who have no fear. They are very rare, but Campbell was one of them."

The decimated RAF units continued to operate from Sumatra until the Japanese dropped more than 200 parachute troops around Palembang. The defending planes and pilots drew back to Java, except for Art Donahue and a few others who, having found Sumatra's primary P-1 airstrips untenable, moved into a temporary base, P-2, near Palembang. Their outlook worsened February 15 with the surrender of Singapore, concluding a hopeless struggle that cost the Allies some 9,000 dead and 130,000 captured.

After the Japanese paratroop landings at P-1, Donahue, on a patrol mission near Palembang, observed antiaircraft fire from barges moving up the Moesi River and radioed his fellow pilots to hold off until he had suppressed the guns. On his first pass over the barges he was struck in the leg by a 20 mm shell. He stanched the wound with a glove and continued attacking the barges until his ammunition was exhausted. This heroic action won him Britain's Distinguished Flying Cross.

[Donahue received treatment at a Dutch hospital in Bandoeng, Java, and then was sent by hospital ship to Ceylon for further care. In 1941 he had written several magazine stories on his combat experiences which became a book, *Tally-Ho! Yankee In a Spitfire*, and while recuperating in tropical Ceylon (Sri Lanka), he wrote a second book, *Last Flight from Singapore*. Recovering his health, Donahue rejoined his old unit in England as squadron leader. On a reconnaissance flight September 11, 1942, he reported by radio that he had shot down a Junkers 88 and that his engine was overheating and he would have to ditch in the English Channel. No trace of him or his airplane was found.]

The 15 RAF pilots remaining in Java on February 23 were called to a meeting in Batavia's *Hôtel des Indes* and told that the squadron had taken such a beating, and casualties had been so high, that Headquarters had decided to move nine of them out. Six would remain in Java to fly the four Hurricanes that remained serviceable.

New Zealander Harry Dobbyn would lead the last-ditch unit, a squadron in name only. The squadron leader called for volunteers for the five other posts. Campbell and a New Zealander sergeant pilot, John A. Vibert, responded. The three remaining slots were filled by drawings from a deck of cards, with holders of the lowest cards taking the assignments. Campbell recalled:

> We had told the people in command that we did not have enough planes. We told them, "We have enough experience, we know the Jap tactics, we know we can beat them at their own game if given half a chance." But we had our orders.

We were told that Headquarters wanted the Japanese to land on every island on the chain down to Australia, to slow their advance and give the Australians a chance to build up their defenses. "Attack everything that comes in, no matter what, and force it to land on Java," we were told. "If you do that you will have done your job."

Churchill had said that the RAF would fight to the last aircraft and then take to the hills and fight to the last man. That was us!

Five days later, on February 28, the four Hurricanes took off to intercept two formations of Japanese bombers accompanied by 30 or more fighters. The one-sided battle was brief. Campbell managed to shoot down one of the Zeros before his own plane lost half of its right wing. He forced his way out through the damaged canopy and parachuted into a rice paddy. He was bleeding from deep body scratches and from shrapnel wounds. His left arm was partially paralyzed, and one leg had been twisted severely.

Some Javanese found him, fed him, and turned him over to Chinese in a boat, who transported him down a river to a Dutch patrol boat. The patrol officer had already found the wreckage of one of the other Hurricanes, with a body he identified as that of Harry Dobbyn. Dutch authorities returned Campbell to Batavia, where he learned that the former squadron now consisted of only four pilots—Vibert and three Englishmen—and two Hurricanes. The Japanese had landed close to Batavia March 1 and were advancing on the city.

In effect, this was the end. "The Dutch in Java, upon whom we depended for ground support, never fought a major battle there, and they refused to fight now," Campbell said. "They surrendered the island and declared Batavia an open city. Some of them even stole the two Lockheed Lodestars that were supposed to be our transportation and used them to fly their wives and children out to Australia. Those of us who were left had to take to the hills and exist like guerrillas for the next three weeks."

At one Javanese town that the Japanese had not yet taken over, the grounded pilots and their remaining crew members appealed to the Dutch officer in command for help. Instead, the fliers were told that the Dutch had surrendered in order to prevent suffering for their families, and the airmen must turn in their weapons also. The fliers refused to comply and escaped in the truck they had acquired. In the resulting exchange of gunfire, one of their members was killed.

> The British army men with us started raiding for food. We were on the south coast of Java, and soon a Japanese patrol picked us up. One of our men started to run and was shot dead. The rest of us surrendered. The date was March 20, 1942.

Campbell remained a prisoner of the Japanese in Java for three and one-half years, until September 14, 1945. He said that, because he had developed typhoid fever, he never was moved off the island. By the end of the war he was a six-foot skeleton weighing 100 pounds.

At one point Campbell and three other prisoners plotted to steal a Japanese twin-engine plane parked near the camp and fly it to Australia. The night before the planned escape, however, Campbell reported to his companions that through a window he had watched the Japanese testing the plane. One propeller was not turning. Campbell said it seemed obvious that the plane was not yet fit to fly. He urged that the escape attempt be postponed.

> The three other men decided to go without me. The Japanese caught them as they tried to start the plane, and they killed all three.

Another group of American civilians and pilots actually found and repaired a Dutch plane in Java, after the Dutch surrender, and got it into the air, taking their crew chief along as a passenger. After getting safely off the ground, they found that they could not raise the landing gear. The crew chief had forgotten to take out the pins that held the gear in the down position.

Campbell said the men flew around for a time, trying to make up their minds about what to do.

> They saw a light and, thinking it was another low-flying plane, tried to get away from it for a while. Finally they figured out that it was Venus. In the end, knowing they never could get to Australia or some other safe place with the dragging of a lowered landing gear and in view of the fuel they had used up trying to avoid Venus, they had to land back in Java again.

Campbell recalled that, during the long prison years, the surreptitious reception of radio broadcasts was a tremendous morale booster. "The Japanese even used to conduct body searches for the radios they suspected we had concealed," he said.

> Once an Aussie named McNally got by with an exceptionally gutsy trick. Usually we had the radio buried in the ground in a tin box, where it couldn't be seen. This time we had been caught by surprise, and the radio was lying in the open, under a bed roll.
> After he had been body-searched, McNally whispered to me to try and keep him hidden a bit. Then he stepped back, picked up the radio, and put it in some clothing he had tucked under his arm. Since he was now among the prisoners who had already been searched, the Japanese did not notice the radio in spite of its bulk, and he was not searched again.
> During a period when our radio was not working, I came to suspect that a group of Dutch pilots had one. There was a technique to finding out where the camp radio was. It got so that you could sense it from the comments of the men in the vicinity—slips of the tongue, occasional mention of what sounded like fresh news. I made a point of checking in with one of the Dutch pilots I knew and suspected of having a radio, at an hour when it was customary to pick up broadcasts. Finally he said to me, "Well, you might as well know," and they let me listen in on the reports.

Campbell said that although the basic purpose of having certain persons monitor radio news broadcasts as regularly and frequently as possible was to keep all the prisoners informed on developments in the war and particularly in southeast Asia, "We quickly learned that the news we gathered had to be restricted, for the safety of the POWs in general."

If too many people learned about up-to-date news happenings too quickly, inevitably they would talk, and the Japanese would overhear conversations indicating that these prisoners were extremely well informed. Then they would track news sources down and seize our equipment, and execute the operators.

It got so bad that we had to hold up news reports, sometimes for weeks, until comments by the Japanese guards made it obvious that they were familiar with the more recent developments. Then, when we passed this sometimes outdated but corrected information to the POWs and they started to talk, the Japanese did not pay much attention since this was old news—stuff that would come out gradually anyway just with the passage of time.

The worst blabbermouths were the Dutch. We learned to keep everything from the Dutch POWs for the longest periods so that there would be the least amount of suspicion.

Campbell said Japanese treatment of the prisoners was harsh. "Some of our people in the camp were put in cages for punishment," he said. "We lost 75 percent of the people taken prisoner, due to disease, starvation or brutality."

At the same time, Japanese army discipline on its own troops was rough. I saw an officer beat a Japanese enlisted man with his metal sword carrier almost to death, in the name of discipline. Another did beat a Jap soldier to death with a heavy bamboo club.

The Japanese could never understand why we complained. They said we were treated no worse than any others. They fed us enough food to keep us active, because they wanted us for working parties. Some of the parties sent to the outer islands never got back.

What kept me alive was a burning desire to live long enough to kill a Jap and skin him alive. I mean that literally. Hate will do that. But looking back, I must say they never treated us worse than they treated their own people. A Korean guard treated us worse than the Japanese ever did. Finally my hatred died, and I accepted the fact that the Japanese had to do what they did.

Campbell said he was considered dead by the RAF for 18 months after becoming a Japanese prisoner because of the report brought back to England by Art Donahue that he had been shot down and probably had not survived. Campbell finally managed to get word out to family and friends that he was in the concentration camp in Java.

After V-J Day, the RAF flew Campbell to London, by way of Calcutta, for hospital treatment. The American went into London on a pass, in part to check up on his old hangout, the Eagle Club. Things were much the same. Mrs. Dexter, dear friend of all the Eagles, was still running the place.

"Red, why aren't you wearing your DFC?" Mrs. Dexter asked. Noting his startled look, she laughed. "Oh, you don't know about it. Of course not, you've been in prison. It was in the papers a long time ago. You must go to the Air Ministry and claim it."

Campbell learned to his surprise that Air Commodore Vincent, a World War I ace and RAF group captain and Hurricane station commander who had shot down four enemy aircraft on a single mission, had recommended him for the Distinguished Flying Cross.

> I remembered having one chance meeting with Vincent in which I was insolent toward him. He had said something about a Hurricane flying straight and level with a Zero after him and called it a "damn poor show, to run away," and that irritated me. I told him, "You know nothing about fighting Zeros. That Hurricane pilot was me." And I stormed out.

Campbell went to the RAF and found that the award had been published in the official gazette. He picked up the medal in October 1945. "Since I was considered dead when it was first announced, I called this my posthumous award," Campbell said.

6

Action and Revenge in North Africa

LIKE RED CAMPBELL, Don Geffene, and Art Donahue, who had preceded them to Egypt and points east, more Eagles, searching for action, headed for the desert war in North Africa in 1942. They soon were fully engaged in battle. Many were shot down, and few of those survived. Yet, one caged Eagle sprang free, bringing his own form of revenge to the advantage of the Allies.

In April 1942, Jim Griffin of Syracuse, New York, and three other members of 121 Squadron—Bradley Smith of Yonkers, New York, Cliff Thorpe of Beverly, Ohio, and Norman Chap of Chicago—restless over the slow pace of the air war in Britain, won RAF transfers to the Middle East. By ship to Freetown, Sierra Leone, Takoradi, Gold Coast, and Lagos, Nigeria, and by DC3 aircraft across North Africa, they made their way to Cairo, Egypt, where they were joined by other former Eagles, including Edwin "Bix" Bicksler, recently of 133 Eagle squadron. Chap was still a sergeant pilot, but the others by now were pilot officers. Griffin recalled his impressions of their adventures in North Africa:

> I'll never forget our first arrival over Cairo, late at night, as we circled for our approach to Heliopolis airport. There was a full moon over the desert, 10 times larger than any that we could remember seeing in America or England. Cairo had been declared an open city, and all the lights were on—a beautiful sight, something like one of the Arabian nights you read about as a kid in school. A big change from London, where we hadn't seen lights at night for a couple of years. I don't believe Cairo was ever bombed, in spite of all those lights.
> Underneath the moonlight, though, Cairo in war-time was a dirty, filthy city. I don't think many Americans ever really cared for the place. The

people were unfriendly, and perhaps that attitude led to our natural dislike of the country.

We were warned about drinking their water, of course, but no one cautioned us about their beer or their soda. We called Cairo beer "onion beer." Every bottle you opened tasted different. The same thing was true of their soda water. I drank some lemon soda and could still taste it an hour later.

We were staying at the Continental Hotel, and I woke up in the middle of the night with a raging fever. It turned out to be amoebic dysentery—a severe type of guppo gut, where you passed blood. It was probably the most horrible illness I ever had. To get to a bathroom you had to walk down the hall 50 or 60 feet. I crawled the distance and lay on the tile floor, burning with fever and desperate with thirst. Each time I got up and took a sip of water I would immediately vomit. Don Hall, my roommate, called an RAF doctor, but Rommel's advance was on and the hospitals were all full. I lay in my hotel bed for about 10 days.

Another time in a Cairo hotel I scrambled out of bed with every inch of my body covered with welts, and an itch so bad it almost drove me crazy. Fortunately, this time I was able to get into a hospital. They smeared me with a white ointment that didn't do very much for hours, but finally the rash went away by itself the next day or two. This was an African disease that was very uncomfortable and very painful.

I've had bad things to say about Cairo, but there were good things too, one of the greatest being their dish called prawns and rice American—small prawns steamed in tomato sauce, covered with steaming rice. Bix, Don Hall, Chap and I used to love them. We'd go into a restaurant and each of us would order double portions. There was no way we could each consume two heaping platefuls, but after being in food-rationed England for so long we'd just gorge ourselves. This was one of our pleasantest memories.

The first time I took a Hurricane out from our station in the desert near Cairo, I wasn't wearing goggles. As I gunned the engines and started to take off, air currents inside the cockpit swept up all the sand from the outside, and for a few seconds I was absolutely blinded with sand in my eyes. We lived in tents along the road between Cairo and Alexandria, and at 3 P.M. every day, as regular as the sun setting or rising, a windstorm would come up. If you walked into the desert for as much as 100 yards, you had to be sure of your direction back to camp because everything would be obscured a foot in front of your face. It was worse than the worst fog in London.

Air raid sirens would blare into your ears almost every night, and before you were completely awake you would be diving into a slit trench or bomb shelter. The Germans usually wouldn't pay much attention to the fighter squadrons, and instead concentrated their attacks on the bomber squadrons on the far side of the road from us. A minor inconvenience was that in a slit trench you not only had to be afraid of snakes—puff adders or whatever vipers were in that area—but scorpions as well. Almost every rock you picked up or kicked over during a day would have a family of scorpions under it.

Griffin recalled with particular fondness Bix Bicksler, a husky, tough ex-football player, the son of a minister in Oak Hill, Ohio, "one of the wildest—and nicest—men I ever met in the RAF or the RCAF."

Bix, who had been in 133 Squadron, was shipped out for getting drunk one night and trying to blow up the mess hall with a charge of explosives. On the way to the Middle East he and an Englishman who had commanded a fighter wing drank too much and sprayed the stateroom wall above a sleeping friend as a prank. Luckily their intentionally high aim was good, and the friend did not try to jump out of his bunk while the bullets were still flying. Bix normally was quite well behaved, but when he got drunk or something set him off, he could become absolutely wild.

One evening Bix and some of his friends wandered into a bar in an area of the city where there had been trouble and people had been set upon by gangs and killed. Later he left the bar and sort of wandered off a little by himself. He was attacked by six Arabs carrying two-by-fours and other clubs. They all jumped him, but that was a tactical error. Six-to-one odds were not enough for fooling around with someone like Bicksler. Bix was unarmed, but he grabbed one of the two-by-fours away and began beating the guy with it. When the other pilots rushed out of the bar they saw the Egyptian lying on the ground, bleeding and unconscious, and Bix was chasing five Arabs down the street. His friends had to call him back to stop him.

The Cairo paper reported the next day that an Egyptian had been found dead in that location, his skull crushed. The slaying might have been considered a form of vengeance. Quite a few British soldiers and RAF people had been killed by Arab gangs. Most of us in that part of Egypt came to look upon the Arabs as more of an enemy than the Germans or the Italians.

Bix transferred over to one of the American squadrons in the desert. I saw him once afterward somewhere, and the only part of his U.S. Air Forces uniform he was wearing was the hat. He had on RAF wings and tattered shirt and pants. He said he spent his uniform allowance to get drunk.

One night, a couple of months after Bicksler's transfer, Don Hall and I were sitting in a rooftop bar in Cairo, drinking beer and watching couples dancing. A fellow we knew, a pilot, came over and said, "Did you hear about Bix?" This was the day after the Palm Sunday massacre of April 18, 1943, in Tunisia. Right away, before another word was said, I knew that Bix had bought it.

I could hear the pilot talking, telling us that Bix had been shot down and was missing, but I could not speak. I don't think I have ever been so shocked in my life. This was one of my best friends. I remembered how we used to sit and discuss what we were going to do after the war, start a flying school or something.

Luckily for me Don was there, and he kept feeding me beer. Finally I got numb. Don and I went out and got in a fight with a couple of British army officers. I lost one grade in rank—flying officer came automatically—when I was called up before the Air Vice Marshal, and he read me up and down and took my rank away from me. Then, after all the official business was over, he sat me down, gave me a drink of wine, and we talked it over. He had been a boxer, an amateur champion of some sort, in England.

Action and Revenge in North Africa   55

One of the men we had been fighting with was a young army officer, a Welshman, who had cursed not only all the RAF but all Englishmen as well. The AVM—or air commodore, I have forgotten the rank—patted me on the shoulder and told me to stay out of trouble. In those days, this was like telling a kid in a candy store not to eat any candy. Something was always coming up.

Edwin Bicksler and Norman Chap had become two of my best friends in the service. I was still feeling the shock of hearing about Bix when, walking down a street in Alexandria while on leave, I met a Canadian flight lieutenant who had served in the same RCAF squadron with Chap. He stopped me on a corner, said he knew that Norm and I had been close friends, and he wanted to tell me what happened to Chap.

This was somewhat unusual because normally we all tried to forget what happened yesterday. In a sense, though, it was as it is in a family or with really close friends—they should be told of important events. The Canadian obviously didn't like remembering this himself.

The flight lieutenant said he and Chap had been scrambled and vectored to intercept bandits headed toward Allied lines. They gained some altitude and had just passed the front line when they saw the wave of enemy bombers protected by 50-plus fighters. There wasn't much two Hurricanes could do against 70 or 80 enemy planes, but they dove at the lead bomber, firing as they went. By a remarkable stroke of luck they managed to break up the enemy formation. It was incredible that two fighters could take on that many aircraft and cause so much confusion that the bombers would mill about and head for home without dropping a bomb.

The Canadian said trouble really began once they were through the bomber formation and had to fight their way through the swarm of Messerschmitts. He and Chap dove down to within a few feet over the deck and started weaving their way back toward the front lines. The Canadian said he saw an Me just ahead, turned into it, gave it a burst, and saw it hit the ground. He looked over and saw Chap, slightly behind him, chasing another Me off to one side. While Chap was firing at that target, another Me was closing on his tail. The flight lieutenant screamed over the R/T for Chap to break, and at the same time tried to turn back into the line of combat to get at the second Messerschmitt. Chap just kept on firing and all of a sudden the second Nazi plane got in range, gave a burst, and Norm went into the ground.

Norm Chap was killed at Sidi Barrani, Egypt, November 7, 1942, five months before Bix's death, but somehow this news was slow reaching us. Chap was one of the finest gentlemen of all the Eagles. He drank, as many of us did, but unlike the rest of us the more he drank, the quieter and more reserved and gentle he became.

The Canadian flight lieutenant also considered Chap one of his best friends. He said one of the most satisfying achievements of his life was that of being able to slow down sufficiently to turn into the Me that hit Chap and destroy it.

The Canadian did not elaborate as to how he escaped from the enemy fighter swarm and got back to base. He simply told me about how his friend, and mine, had died.

Oddly enough, after many years I can plainly remember standing on the street corner of the Egyptian city and listening to him, but I cannot remember his face or his name.

Sadly for Jim Griffin, his closest Eagle compatriots in Egypt were now all out of action. On September 15, 1942, Cliff Thorpe had been reported shot down and held as a prisoner of war.

* * * *

A remarkable Eagle was Harold Fesler Marting—remarkable both for survival of his capture-and-escape experiences in Egypt, Libya, Greece, and Turkey, and for the storytelling skill of his diary of that time of his life. His novelist sister Lennie gave him the diary in 1940 when he enlisted in the Royal Canadian Air Force and wrote on the flyleaf, "Buddy, you'll be glad later that you wrote it down, so please do. And some day, I want to have my share in telling it." Buddy Marting followed her advice, and Ruth Lenore "Lennie" Marting—Mrs. Don Silvers—saw her wish fulfilled.

Indiana-born Marting, known to his friends as Hal, left high school to join the Marines, and later served in Haiti during the uprising of 1929. Returning to civilian life, he learned to fly.

Joining the RCAF in October, 1940, he took flying training in Canada and England, and was assigned in August 1941, to the Royal Air Force 121 Eagle Squadron at Kirton-in-Lindsey. Action initially was dull, particularly when heavy autumn weather settled over the British Isles and much of Europe. In early November he wrote in his diary, "Have asked very emphatically to be transferred out of this squadron and am sure I'll get it this time . . . . Am damn well fed up with this squadron and this station."

Eleven days later he was a member of 71, the original Eagle squadron, at North Weald. Within days he was complaining in his diary, "Weather is terrible here . . . . No flying today, thick fog. Am fed up." In February 1942, he requested posting to the East. He arrived in Durban, South Africa, in May, and the following month left by ship for Cairo with three other former Eagles—Eddie Miluck, Mike Kelly, and Wally Tribken—who had hopes of joining General Chennault's already-famous Flying Tiger volunteers in the China-Burma-India theater.

Marting quickly found himself assigned to 450, an Australian squadron flying Kittyhawk dive-bombers. On August 2, 1942, in his first sampling of desert operations—an air assault with 500-pound bombs against columns of German tanks and trucks—he shot down a Messerschmitt 109. He claimed another enemy plane October 20. The RAF, meanwhile, had informed the pilots that the Allies now had achieved four-to-one superiority in troops and five-to-three in planes over the enemy, and were about to push the Germans and Italians "right out of Africa."

Marting earlier had visited the American legation in Cairo to apply for a transfer into the U.S. Army Air Forces, and had also inquired about a possible U.S. Navy air post. Now, however, with excitement in the air, he wrote in his diary for October 22, 1942: "The push starts tomorrow night. There will be no rest for us for some time now. This coming battle is intended to be the turning point of the war, so I am proud to be in the main striking force."

The push just ahead, the "coming battle," turned out indeed to be a major advance for the Allies in the Middle East, one that has come to be known as the Battle of El Alamein. During the engagement starting October 23 at El Alamein—the Mediterranean seaport about 60 miles southwest of Alexandria and only 150 miles from British-held Cairo itself—and continuing for several days, the powerful Nazi advance spearheaded by General Erwin Rommel was halted and turned back, and German and Italian forces were hammered back into Libya. Marting was not to learn of the British triumph for some time to come, however. This was to be the point of his departure from combat. His diary account follows:

> 10–23–42. Off at 0700 to dive bomb El Daba airdrome. We went by the sea route with 450 on top and me leading 450's top cover of six. We were last to bomb and dived at about 75 degrees. I could see many good hits by the others before me.
>
> We had seen no enemy aircraft in the sky. I aimed my bomb at two planes on the west end of the drome and released it at about 6,000. I made no attempt to see where it landed.
>
> I pulled back up towards our formation at a very steep angle and saw it about half a mile ahead. At this time four other Kittyhawks went by me, which I presume were the outer sections of my top cover.
>
> I realized I was much too far back from the formation but there was nothing I could do about it. I had been flat out since starting the climb after bombing. I was getting full RPM and had about 45 inches boost, but could not keep up. I could not see anyone behind me, so apparently my No. 2 had lost me.
>
> When I was at 10,000 I saw four Me 109s about 1,000 feet below and climbing in the same direction I was going. They were slightly ahead and apparently had not seen me. There was very little I could do, so I attacked, diving on the nearest one, and gave him a two-second burst from about 200 yards. This squirt hit him very well, and a great deal of smoke came out of his engine, and he went into a rolling dive with fire starting on the port side. I believe he was out of control and claim it as probably destroyed.
>
> I could not follow him, as I was naturally worried about the other three, which were sure to have seen the attack. I turned towards the nearest one, which was on my starboard side at my level, and opened fire from his front quarter at about 300 yards. I allowed too much deflection. My tracer went in front of him by a few feet, and he pulled up into a climbing turn away from me. This was only a short squirt, about a second, but I think two or three of my guns stopped then.
>
> I corrected as quickly as possible and started another squirt from his port rear quarter at about 250 yards, but only two guns opened fire. After only a few rounds one of them quit. This threw me well off the target, so

I had to break off. It was hopeless for me to try to catch the formation, so I broke off in a vertical dive and had a good look for the other 109s, which I could not see. I have no idea where they went to. As far as I know, none had fired at me.

On the way down I saw that I was five or seven miles south of the coast and, I thought, opposite the heavy flak area. I decided to try to get back without going out to sea.

I gradually pulled out of the dive with about 450 mph on the clock and leveled off at about 20 feet above the ground. For several minutes there were no shots fired at me that I could see, and my speed slowed down to about 250, which it held. I got down closer to the ground but had to pull back up because it was difficult to see the sand hills in the shadows made by the clouds.

I crossed a large group of dispersed motor vehicles at about 30 feet and was weaving as violently as possible because they threw a lot of light gunfire at me. Nevertheless my right wing was holed twice and one had hit the tail, somewhere.

I continued weaving and could see that I was crossing a heavily defended place by the dug-in guns and trenches. A lot of small stuff was still coming up, and now there was Breda 20 mm too. All together, I was hit five or six times by the small stuff, which did no damage, but finally an explosive shell hit the engine and filled the cockpit with glycol, smoke and cordite fumes. The engine quit, and I had difficulty seeing for several seconds.

I turned off the petrol and the magneto switches and decided to keep heading east as long as possible. I quit weaving to try to extend my glide, and tightened my straps and opened the hood. I decided against flaps, partly because they would cut down my speed and also because I wasn't sure they would work. I had very little time to do anything anyway.

The plane finally touched down at about 100 mph, and it was pretty rough going for a few seconds. The hood came forward and caught my left hand as I was bracing myself, but it only bruised my fingers. I got out of my straps as quickly as possible and destroyed the IFF before jumping out. I got my code card out of my Mae West and tore it into small pieces and was trying to place myself. I was a little dazed, or I would have realized that I was on the edge of their front lines. The tail of the plane had actually hit the barbed wire in front of their minefields, through which 10 or 12 Italians were coming towards me yelling.

Had I not been somewhat dazed I could have had a good chance of running, and I'm sure I would have tried had I been sure that this was the very edge of their lines. However, I made no such attempt, and the Italians came up and took me prisoner. They relieved me of my knife, pistol, belt and watch. I protested violently at the taking of my watch, but one made businesslike signs of shooting me so I gave it to him. Later protests to officers were of no avail. Although they made a few pretenses of trying to recover it for me, I did not get it back.

The Italians took my Mae West, helmet, parachute and the cushions out of the plane, presumably for souvenirs. We went back through the minefield and were met by an Italian officer and by a German orderly who acted as interpreter. I was still quite nervous, and they gave me half a cup of anisette, which made me feel much better.

We chatted and smoked for about half an hour. Then I was put on the back of a motorcycle behind an Italian and we started for the rear over a very rough track. It was difficult to hang on even at a very low speed, and once we both fell off.

We arrived at a group of dugouts, which turned out to be company headquarters for the 62nd Infantry. I was met by a red-headed officer who was very kind and cordial, and gave me a cup of good coffee. I was placed in the back of a Chevrolet trunk with three soldiers, and with one NCO and driver in front. One of the soldiers made attempts to get my cigarette lighter, but I refused even to let him touch it.

At Brigade headquarters I was questioned by an Italian intelligence officer who spoke a fair amount of English. The only information I gave him was my name, rank, and squadron number—data which was on my identity card in my pocket.

A general came over and spoke a few words to me, out of curiosity, I think. They seemed quite friendly and made no threats. Several remarked that for me the war was finished, and all seemed quite surprised that I was not happy about the whole thing. They tried to paint a glowing picture of what life would be like for me as a prisoner in Italy.

They asked me also about the treatment Italian prisoners of war received. I told them quite truthfully what I had seen of it, and they appeared to believe me and were somewhat relieved.

After about an hour I was driven in a small car to what I supposed was divisional headquarters, where I was questioned again along the same lines. I got quite weary of the questioning here, as I had to stand bareheaded in the sun. Finally they searched me again and drove me a considerable distance to the Italian General Headquarters. The officer who questioned me here was very clever and tried in various ways to trap me into giving information. He got none out of me. He was curious about the effectiveness of their camouflage, and whether we knew the location of their HQ. I refused to answer such questions as how I arrived in Egypt, by ship or by plane; how long it took our mail to reach us; how many planes in each squadron; where I took my flight training. He asked whether I intended to escape, and I said I would if there was any opportunity. "That would be very dangerous," he said. When I laughed, he added, "You are a good officer."

I was taken to another dugout, where a colonel asked why America made war on Italy. I replied that it was because Italy and Germany had taken the homes and liberty of millions of people in Europe.

"Africa is no place for Englishmen and Americans," he said. "That is true," I replied. "Africa is the place for Africans, not Italians."

This made him furious. "The Italians and Germans will teach you a lesson," he said.

"They already have, in countries like Poland and Greece," I said. I thought he would burst, but he stopped questioning me.

I was driven to a small prison camp which had no shelter of any kind, and must have been merely a place to hold prisoners in transit. My Italian guards turned me over to two German officers and received a receipt acknowledging the transfer.

There was only one prisoner in the fenced-off enclosure, a South African negro. The Germans did not put me in the enclosure, but had me sit in front of the guards' hut. They gave me the same food they were eating—soup and bread. They were very friendly and showed me pictures of their wives and girlfriends.

After dark I was driven to Luftwaffe headquarters on the beach near El Daba airdrome, where they had a transit camp for captured airmen. I believe I could have escaped during this trip, with a little luck, but I was too tired and did not have the energy. It would have been risky and [the trip] needed the best I could have had.

Two officers took over my questioning. One, a young fellow, spoke very good English but overdid the friendship act considerably in trying to put me off my guard. They tried to wheedle all sorts of things out of me.

Obviously trying to find out how long I had been in Africa, they asked if I knew pilots in the squadron who had been taken prisoner before me. In this way I learned that Weing, Lindsay, Evans and Holloway were all prisoners. I was careful not to commit myself when they mentioned the names.

When I refused to answer questions about our forces, they laughed and said I might as well talk because they already knew all the answers.

"How is your new squadron leader Williams getting along?" one asked. That rather set me back on my heels because he had been squadron leader only three days. "If you know all the answers, there is no need to question me," I said.

The officer in charge of the prison gave me some liverwurst sandwiches and a glass of whiskey, and then took me into the prison itself and put me in a tent with three other Americans. They were already in bed and were suspicious of me, so we did not talk much.

10–24–42. I awakened stiff and sore from sleeping on a straw mattress about two inches thick, spread on the ground and alive with sand fleas. We each had only one blanket and slept in our clothes, only removing our boots, which were taken away for the night by the guards.

We breakfasted on ersatz coffee and bread, and I got acquainted with the other fellows in the tent. They were the crew of a B-25 which had been shot down by flak two days earlier. The two gunners had bailed out and were not yet heard from. The crew in the tent were Jim Cleary of Columbus, Ohio, pilot; Francis Finnegan, Buffalo, New York, copilot; and Bill O'Berg, Iron River, Michigan, the navigator, from the USAAF 83rd Bombardment Squadron.

In the other three tents there were two Englishmen from a Wellington bomber, both badly cut and burned; Pilot Officer Hogg of 113 Squadron, a Canadian whom I knew by sight; and from the South African Air Force, Lt. Mackay and Sgt. Corson. Corson was gunner in a Boston hit by flak. He bailed out, thinking the plane was out of control, and was greatly surprised to see it recover and continue on its way, leaving him hanging in a parachute.

During the morning Mackay, Hogg and the two Englishmen were put on a truck and sent off to Tobruk. Before they left Mackay told me that another flier, named Evans, had left a map and two saws buried in our tent. We spent the morning looking for them and finally found all three.

We talked escape most of the time but it looked impossible from this camp, since it was right in the middle of a very large airdrome area. None of us could speak German, although Corson could understand it perfectly and spoke Afrikans, which is really Dutch. We more or less abandoned ideas of escape from the camp, and hoped for an opportunity in transit.

During the afternoon they brought in Sgt. McMinn of Washington state, who was one of the gunners on Cleary's B-25. They were glad to get together again. The other gunner was in hospital with a bullet through the forearm.

I was taken to the tent of the officer in charge and interviewed by Count Eckstadt, who professed to be a newspaper correspondent. He spoke good English and told me he had worked three years in the States for Borden milk company as a research chemist. He said he had met Ford and Edison.

He tried to get me to make a recording of my experience to be put on the radio to the States. I refused because I thought they might use it for some of their propaganda. He was a quite nice fellow and we talked nearly an hour on world politics.

He asked me if I thought England and the States would try to take revenge on the German people if Germany lost the war. I said no, but that we would have a tough time trying to keep the Poles, Czechs, Dutch and all the rest of Europe from taking theirs.

He admitted that he thought Germany had bitten off more than they could chew. I told him there wasn't any possibility of their being able to lick the rest of the world, and that they were stupid to try. He rather agreed with me but said that now they would fight to the end because they were afraid of the aftermath. He added that there were no shortages of anything vitally important yet in Germany and thought the people were not near to collapse.

10-25-42. They woke us before daylight and we were piled onto a truck with five guards for the trip to Tobruk. They gave me a French overcoat to wear, like those already given to the others. They also gave us rations for three days: several loaves of bread, tins of bully beef, potatoes, pork and jam. Except for the bread it was all English stuff that had been taken when Tobruk fell. The Germans laughingly told us they had captured enough food there for more than six months' need for their forces in Africa.

The six of us prisoners sat in the back of the truckbed with our guards in the front, facing us. There were too many of them for us to hope to overpower. We knew they had orders to shoot without warning.

We arrived at Mersa Matruh about 0930 and stopped for breakfast. Leaving Matruh we were joined by two very arrogant officers and another soldier, which crowded the truck quite a bit and made escape more remote than ever.

After sundown we stopped for half an hour to wait for the moon to come up. No lights were allowed; we had to have moonlight before we could go ahead. During this wait it might have been possible for some of us to have made a break, but we were in between minefields with no maps, food or place to hide. My map, which I had tied around my leg just below my left knee, had come loose during the rough ride, and one of the officers had taken it from me.

We arrived at a stopping place called Gambut and the six of us were put for the night into one room with four of our guards. We slept on a blanket each, on the bare floor, and were quite chilled before morning.

It may have been possible to escape at this time, but there still remained the problems of food and water and maps. None of us seriously considered breaking out, but we did talk of the possibility of stealing the plane that was to take us to Greece. We decided to seat ourselves in the plane in such a way as to be able to take advantage of any opportunity that might arise.

10-26-42. We pushed on before daylight and finally stopped at the edge of Tobruk for breakfast. Going down the hill towards the town later we saw an airdrome, a tiny harbor containing 10 or 12 ships, and heaps of stone that had been buildings.

As we arrived at the airdrome, German Junkers 52s began taking off in formations of about 20 at a time. At one point I counted 60 planes in the air. They stayed in vee formations and flew north at about 100 feet. We had lunch right there in the sand, and it was interrupted twice by air raid alarms, although we saw no planes.

All morning all types of aircraft were landing and taking off: Me 109s, Me 110s, He 111s, Ju 52s, Savoias, Bredas and even a couple of gliders towed by Heinkel 111s. Most of the 109s were coming in from Italy—replacements with long-range tanks under the belly.

About 1 P.M. 30 Ju 52 transports came in, and we boarded one. There were no seats. The baggage had been piled along the sides for us to sit on. The crew consisted of pilot, copilot, wireless operator, rear gunner and one man in our compartment who looked after the baggage and seating. In addition to the six of us prisoners, there were six guards, one captain and two German soldiers. All of the Germans were going home on the month's leave granted them after six months of service in Africa.

Three of the guards were corporals, one of whom spoke fairly fluent English. His name was Fritz Eiben and his home was on one of the small islands off the northwest coast of Germany. A handsome schoolteacher, 20 years old, he was well educated and a nice boy, and we all got along very well. Fritz was considerate of our needs, and we were grateful for his presence many times.

We had some lengthy discussions about the war and about German politics before the war. I was somewhat surprised to learn that all of these soldiers still accepted the official German communiques as gospel truth. They believed our governments misled us about our losses in shipping and planes.

I told Fritz it was impossible for the RAF to publish false figures on plane losses. He was only slightly impressed. He said he knew we had lost 62 planes in Africa October 22, and I knew for a fact that we lost only 11. He also believed Roosevelt to be a Jew and a dictator. We laughed so hard at this that he later felt a little silly about it, but he still wouldn't believe it would be possible for an American to denounce the President publicly without being thrown in jail. He said he was sure that only a small minority of Americans wanted to declare war when we did. He argued that Roosevelt had done this himself and there was nothing the American people could do about it.

Fritz interpreted our remarks, and it made me furious to see the other guards look at us and shake their heads in pity, as if to say "you poor misguided fools." That helped me a great deal in reaching a decision to escape, but what made me even more determined was their total lack of feeling for the peoples they had defeated and were robbing to feed their war machine.

The poor Greeks, Poles, Dutch, and other people of Europe were just so many animals to them. Their feelings for the Russians were a mixture of hate and fear. They despised the Reds and could not understand why Russia should keep on fighting and losing so many men. All of the Germans were frightened at any prospect of being sent to the Russian front.

We were so scattered about, inside the plane, and so well guarded and so greatly outnumbered that there was no hope of commandeering the aircraft. In addition, we were flying in close formation with 29 other German planes. We refueled in 20 minutes at a small airdrome on the north coast of Crete, and were on our way again—to Athens, we were told.

I felt sure that Greece would be the best place for a break. I was confident the Greek people would be friendly and would help us if possible. The other fellows were pessimistic except for Finnegan.

Cleary was talking all kinds of nonsense. I believe his crackup had shaken him badly. O'Berg was a close friend of Cleary and would follow in anything Cleary did. I could not count on either of them for much help.

McMinn was calm and collected. I could have counted on him and Finnegan, but both were very inexperienced and had few ideas of any value for what we must do. Corson was a big husky fellow who talked loud and appeared to be just the kind of man we needed, especially since he could understand and speak passable German. I was only slightly dubious of him because of the fact that he had bailed out just because his plane had been hit once. These doubts proved correct later on.

We landed at Athens airdrome at sunset. During the hour-long wait in a hangar for transportation, I wanted to have a go at a break, but guards gave me no chance. When the truck arrived to take us into town, we were put in the back, near the driver, with our guards behind us, so again there was no hope of escape.

We were taken to a small downtown hotel, the Rex, and told we would spend at least the night. Our guards said we probably would get a train for Germany the following day.

We were assigned four rooms on the third floor, with three of us to a room. The beds were clean and comfortable, and we had running water in each room. The guard Fritz, Finnegan and I were in a room on the back side, off the street. We had baths and some food. I had not shaved or washed since the 22nd, and was really crummy.

After we had cleaned up and eaten, the guards asked us for our parole— our promise that we would not try to escape for the period that we would be in the hotel. We had a short conference and, since we were all weary, agreed to give it for the night. I was for it because I didn't want to try anything until I could see the lay of the land in daylight. We had a comfortable evening, and no guards were posted at all.

10–27–42. After breakfast in our rooms Fritz decided that we might take a short walk, still on parole. We walked six or seven blocks around the town but the guards brought us back because of the huge crowds which gathered and followed us. The people were very friendly and showed us plainly what they thought of the Germans and Italians.

When we got back to our rooms, one of the guards told us we could not get a train for Germany until the next Sunday, and must stay on at the Rex until then. We prisoners had another conference and decided to take back our parole. I was sure we would have a good chance to escape here.

The guards were visibly disappointed and quite worried at this decision. We tried to laugh off their discomfort by saying that it was against our orders to give our parole anyway, and we had done so only so that we could all get some rest the night before. That did not help much. They were still very suspicious.

I was almost sorry for Fritz. He told me that the guard who let a prisoner escape was put in prison for at least two years. No wonder he was worried, because at least 75 percent of the Greeks probably would help us if they could. This night the guards did two-hour watches in the corridor.

10-28-42. The guards decided to keep us all in one room during daylight hours so that a smaller number would watch us and the rest could go about town when they were free. We decided to do everything we could to relieve their suspicions, and to make no breaks until we all were ready to try. From then on we were careful to do what they asked and rarely left the room during the day except to go to the w.c.

The room occupied by Cleary, O'Berg and a guard fronted on the street and had a small balcony where we were allowed to sit. Again, the people on the street made all sorts of friendly signs to us, and several sent us cigarettes. One fellow sent Corson a toothbrush, paste and some razor blades which he had asked for by sign language.

We were greatly affected by the hungry look of the people here. Some of the hotel employees begged us for some food. I slipped a tin of potatoes to an old man and he kissed me. I thought he was going to cry. Two women who cleaned our rooms always ate the scraps and crumbs left after our meals.

10-29-42. We could see that the streets were full of soldiers all day long. There would be little chance of escape without either German uniforms or civilian clothing.

I worked out a plan for all of us to have a try by disposing of our guards one or two at a time, gagging and tying them to the beds and taking their clothes. This would have been extremely hazardous because the slightest slip would have ruined the whole plot. I do believe it could have been done, but am glad we did not try it. All were willing but Corson, who frankly admitted he was afraid to use violence for fear of being shot if recaptured. If we were to go in a body it would be necessary to have someone who spoke German. Corson's refusal threw that out of gear.

I was pretty sure I could get away by myself because Fritz had an extra uniform in his baggage in our room that I might steal. If I could get 30 minutes without my absence being detected, I thought I could get down a fire escape on the rear of the hotel, cross the courtyard there and scale its 12-foot back wall. I had no way of knowing if there was a way out along the rooftops or down the other side of the wall because it was impossible to see where an escape route might lie. I was very much frightened also at the thought of being detected while climbing down the fire escape. I had no doubt that if discovered, I would be shot without warning.

I was extremely disappointed to find that Fritz had gone out and had locked the door to our room and taken the key with him. I was also worried because I was wearing English riding boots that were black and decidedly un-German and had leather heels that made a big racket when I walked. I dreaded climbing down the iron fire escape with them, and I wasn't sure I could scale the courtyard wall without rubber soles. Finnegan generously offered me a pair of crepe-soled shoes, but I didn't take them because my boots would have been uncomfortable for him in prison, and because our changing shoes might make the guards suspicious. I decided to make some attempt the next day, if possible.

10-30-42. I managed to make clear to a man in a room across the street that I wanted some clothes. He indicated he would help and have them waiting for me at midnight, down in the street. This was tempting, but I could not figure out any way to get past our guard in the hall and into the street without being seen. Also, it would be out of the question to try to negotiate the fire escape route at night.

The guards had relaxed somewhat during the day, I guess because they thought it would be impossible to make a get-away in daylight. Fritz took half of his baggage out of our room and put it into the adjoining room which was used by the other guards. This dismayed me because I knew that his extra uniform tunic was in that baggage, leaving only a pair of trousers and his cap still in our room.

Fritz locked our room and went out, but returned in about an hour and unlocked it again because he was staying in for the afternoon. About 3 o'clock I told Finnegan I was going to try it again, as I had discovered that Fritz was taking a nap in the room to which he had moved part of his baggage, and the other guard was in the captain's room listening to the radio.

Finnegan, Cleary and O'Berg agreed to cover for me if I was asked for or missed, by saying I was in the w.c. I walked to the toilet quietly, taking care not to wake Fritz who had left open the door behind which he slept.

Closing the door to the w.c., I sneaked into my own room, locked the door, and as quickly as possible changed into Fritz's extra trousers—the German military kind that buckled at the ankle. I searched his bag for another tunic, but there was none. This bothered me because the Jerries always wore a coat, and I had on my British khaki shirt. I found his Luftwaffe cap and put it on. Then I opened the window and climbed out on the fire escape.

The next half hour was the most frightening and exciting that I would ever want to have—far more terrifying than any fighting had been. Climbing down into the court I thought my heart was going to jump right out of my chest. I was shaking so badly I could hardly control myself.

As I passed the second floor an old lady opened her door onto the fire escape and started to exclaim. I motioned her to be quiet, and said "English." She retreated rapidly, closing her door, and I climbed down, resting at the bottom to quiet my nerves.

There were no sounds from above, but as I started toward the far end of the court someone turned on a light in the first-floor hallway that opened on the court. This scared me half to death, and I retreated back up the fire escape to the second floor. There was a ledge along the top of the wall here but it was very narrow—only about four inches wide, impossible for me to use. I decided to risk the courtyard again, although I knew I could be seen by anyone in my room or in the adjoining room where Fritz was catnapping.

I crossed the court to the far corner where a table leaned on edge against the wall and a small pipe—an electrical conduit—extended up to the top. I stepped up on the table edge and pulled myself up the wall with the aid of the conduit. Just as I reached the top the pipe gave way, fortunately without much noise. I scrambled onto the roof and saw a small garden on the other side. A lady working there looked up in surprise.

"English?" she asked. I nodded, and indicated I wanted to get out to the street. She pointed to the door of a storage shed. I jumped down and went in.

The shed led into a small photo shop. Several people were inside, so I turned back. The lady in the garden went and cleared the people out while I waited nervously, expecting an alarm to be raised at any moment. I had now been away from my room for about 20 minutes.

One man remained in the photo shop, and I asked him for clothing. He shook his head and indicated he could not help me. The shop opened into the rear of another building containing offices and a tailor shop which I entered.

There were three men in the room, and one spoke a little English. I explained my predicament and my need for a coat of some kind. They said they had nothing. As I started out again a man entered the building, looked at me with some curiosity, and started up the stairs. I followed him, and on the second floor stopped him and asked for clothes.

He took me into his office and I managed to make him understand what I wanted. He was French and spoke to me in English. He showed that either he could not or would not help.

I decided to chance the street in the hope of getting away without being questioned about my peculiar uniform. I walked along as normally as possible. None of the civilians paid any attention to me, and as luck had it I did not meet a single soldier. The street was crowded, but I was just fortunate enough to get away with my venture.

I reached a residential district and approached a man and again asked for clothes. He could not speak English, but I made him understand. He refused to help me. I am sure that he and the others feared I was a German.

A young girl standing nearby saw the whole thing. I asked her if she spoke English. She shook her head. I went on toward a large park and thought I might make for the hills north of the city and hide out.

Halfway across the park I rounded a turn and almost walked into a squad of German infantry. I doubled back just in time and saw the girl to whom I had spoken coming toward me with an older lady. They beckoned for me to follow.

I stayed some 10 yards behind them, and they led me to a house and into an inner court, and then into a washroom. Three or four women were there, and they indicated they were sending for someone who could speak English.

Within an hour another woman arrived. She introduced herself as

Litsa, and she understood enough English to find out what I needed, and to believe me and help me. It was dark by this time, and Litsa and the older lady and I walked to an apartment house where their friend Angela lived. Her husband had died recently, and his clothing was available.

Angela took me in and immediately gave me a suit, a shirt, and a hat. While I was changing they telephoned another friend, Mariette, who spoke fluent English. Mariette arrived quickly and I told the complete story to her. She was satisfied that I told the truth, and they agreed that I would stay the night.

10-31-42. I spent the day in Angela's apartment. In the afternoon two very charming sisters, Elli and Lilika, came in from the apartment across the hall. Elli spoke English very well. Their brother had been a pilot in the Greek Air Force and had been killed fighting the Italians and Germans.

My friends decided I should spend this night and the following day with a friend of Litsa's—an artist—Costas. I stayed, locked up in his studio. He gave me books to read, and in the evening he, Litsa, and I had a little feast.

11-1-42. It was decided that I should return to Angela's to meet two men who would try to get me back to Egypt. Litsa, Costas, and I walked back through the dark streets. The two men were waiting for us and, after a short conference, said they would help. They decided I should spend this night at the flat of Mariette, so Angela, Litsa, Costas, and I walked down there.

11-2-42. I stayed in all day, of course, and could watch the Germans and Italians walking up and down by our window. There is a building across the street where the German "mattrossen" girls live.

Mariette got in touch with one of her friends who is in an organization to help people get out of Greece. He came this evening and satisfied himself that I am an American, and he also agreed to help. Elli and Lilika visited me tonight and brought me some candy.

11-3-42. Elli and Lilika and Litsa were in to see me today. It is decided that I am to return to Angela's for a few days. Lilika brought me a pullover sweater she had made for her brother. Today is the anniversary of his death.

11-12-42. I am still at Angela's as I write this, waiting and hoping every day to be leaving soon. Elli and Mariette bring me books to read, and every day Elli and Lilika have come by with fruit or cakes or cigarettes. The "organization" also is providing money for my food and cigarettes. All of these people have been kind beyond words, not hesitating to risk their own lives. I will always think of Angela, Mariette, Lilika, Litsa and Elli, as my five adopted sisters.

The only thing I am sorry about is the fun I am missing in Africa now that the biggest push is on. I am leaving this with Ella to be mailed to you, my sister, after the war just in case I don't get back. I think it is too good a story for you to miss, so I'm taking no chances. I hope I get back to tell you this story myself long before it reaches you, but in any case it is much better to put it all down while it is still fresh in my memory. Love, Harold.

P.S. Am learning to speak a little Greek!

* * * *

At this point, Marting's diary narrative ended. The August, 1943, issue of *The American Magazine* featured an article, "I Escape," by Flying Officer H.F. Marting. Its account of his capture by the Italians, and of his daring escape from German guards, conformed in general to the diary's recital. However, the magazine article, published in wartime and subjected to military censorship, made no mention of the Greek women who sheltered him, nor did it identify the other individuals who helped him along the way.

In the magazine article, Marting told of meeting a man known to him only as Bill, who questioned him thoroughly to make sure he was not a German or Italian agent, and then provided food and shelter. Marting told how Bill led him and several Englishmen and Australians on a gruelling hike across steep mountains, in bitter winter cold, to the seacoast and the small boat that would take the party across the Aegean Sea to Turkey.

As for Marting's diary, it reached his sister in Farmland, Indiana, in the mail late in March 1946, accompanied by a note dated March 12 and signed by Angela A. Melidi. The letter is as follows:

> Perhaps my name is known to you, perhaps not. In any case I feel a holy duty to write these lines and send the journal of your brother Harold Marting, who was captured by Germans and hospitalized by me in Athens after having escaped, as you will learn from the details from the enclosed journal.
>
> Dear Madame, I have already written to him, but unfortunately till now I have not his news and I am very anxious. I will be very pleased if you kindly give his news and write to me if he is in good health because I consider him as my adopted brother since we passed many troubles together during his stay in Athens.
>
> I am very happy to be able to send this journal to you. This was his wish, and please answer to me as soon as you receive this letter.

Lennie Marting Silvers, speaking to a reporter many years later, said Angela Melidi was a widow whose husband had been shot by the Germans for his activities in the Greek underground. She added:

> Angela Melidi hid my brother for 12 days, moving him from place to place—including a coal bin—and giving him her husband's clothes, dyeing his hair and moustache, and providing what food she could for his walk out of Greece. The young Greek girl and another girl knitted him a sleeveless sweater, which I still have.
>
> The "Bill" who led the group out of Greece was a prominent Greek Army officer whose name I do not know. He had gone underground and the Germans were so hot on his trail that he needed to get out of the country for a while.
>
> One day when Harold's feet were so swollen and infected that he could walk no longer, the group left him in the lee of a boulder, "Bill" promising to come back for him. Harold became so cold he crawled up the

mountain to the home of a Greek farmer and his son, who gave him food and told him to build a fire in their shed to get warm.

Two Italian military police saw the smoke coming out from under the eaves, and came to investigate. They looked at his forged papers and laughed, and said,"We know you are not Greek. What are you?" Harold was about to give up hope, and said, "I'm an American."

They laughed even harder at such an absurdity, and said, "Gestapo! We know you are Gestapo!" Harold shouted one of the few German words he knew, "Harrouse," meaning "Get out." They went away.

The next day "Bill" came back with a donkey and two-wheeled cart, in which they rode for several miles before having to return the borrowed transportation to its owner. Later a priest in a Model T Ford picked them up, and they were able to ride until that bad climb over the mountain.

The party of fugitives led by "Bill" arrived in Turkey December 17, 1942. During the crossing from Greece, the hunted men hid under a tarpaulin covering the fishing boat's four-by-six-foot hold.

> We sent one of the natives in the curious crowd on the pier off to notify British authorities of our arrival.
>
> No one who has not been hunted like a wild animal for months could quite understand our feelings as we stood there on free soil. We were ragged, unshaven, and filthy; all of us had bad feet and some of us could hardly stand, but we were free men. There was a big grin on every face.

For five days the men rested in hospitals and guest homes. "We were given the best of care and the finest of foods." The leader of the rescue party departed unannounced. "Bill didn't wait for thanks," Marting reported. "I felt pretty bad at first, but I guess it wasn't really necessary. Wherever he is, he knows."

The group proceeded by train through Turkey, Libya, and Israel into Egypt.

> On New Year's Eve I arrived in Cairo and hobbled down to Shepheard's Hotel. It seemed to me that half the American Army and all the jubilant British Eighth Army were in the lobby. And there, as if fate hadn't already been kind enough, were my pals of the old Eagle Squadron, Wally Tribken and Mike Eddie Miluck. They pounded me until my loose teeth rattled.
>
> "Son of a gun!" yelled Miluck. "Even the Jerries couldn't put up with him."

After discharge from the RCAF, Marting became a civilian pilot, testing engines for the Curtiss Wright Corporation. On September 20, 1943, while flying to Buffalo, New York, to arrange for processing at the Eastman Kodak plant in Rochester of films that he and Miluck shot over the desert during their service in Africa, he crashed and met the death he had avoided so many times before.

Great Britain awarded its Military Cross posthumously to Harold Marting. The citation mentioned in particular that, while in captivity, Marting obtained information concerning a number of RAF, USAAF, and South African Air Force airmen who had been taken prisoner, and later passed this on to the Allied authorities. As Lennie Silvers related:

> We might never really have known why he received the Military Cross if he had not told me in confidence when he returned to the United States.
>
> While Harold was held at Tobruk, he noted the arrival time of Rommel's supply planes, and the direction from which they came. As soon as he could, he sent a message, through underground radio from Greece to RAF headquarters in Cairo, as to when and from which direction the German planes would come, and indicated the flight plan the RAF should take, coming in and out of the sun.
>
> The next day they shot down 67 of Rommel's supply planes, and it was soon after that that Rommel's retreat began.

7

Freedom or Famine

IN THE SPRING AND SUMMER OF 1942 the Eagles' air war in Europe rose to a higher intensity. The Luftwaffe was holding back, and with the rapid increase in British and American aircraft production and the arrival of new fuel supplies from the United States, the RAF was going on the offensive. The Eagles pursued fighter sweeps and bomber escort missions over German-occupied France, as well as low-level "rhubarb" raids on ground targets. In August 1942, they flew support for the commandos at Dieppe and engaged in what one Eagle called "probably the greatest air show the world had yet seen." Inevitably, more Eagles became POWs. One Eagle, an almost double ace, had an impromptu tour of France before returning safely home.

For the Eagles held in Germany, the year brought a change of cages. Danny Daniel recalled that "In January 1942, we were moved to Oflag IXA, Spangenberg castle above the town of Kassel, for a few weeks, and then to Stalag Luft III at Sagan, to East Compound, the only one there at the time. After a month or two, we were transferred to Oflag XXIB at Schubin, a little village in Poland, and here we joined another Eagle prisoner, Nat Maranz. We stayed there almost a year and then were sent back to Stalag Luft III."

Bill Hall, Bill Geiger, and others also moved, spending half the year in a camp in lower Kassel, where they soon came to know hunger.

The tumultuous times had two Eagles out of action before the end of April. William L. C. "Casey" Jones of Parkton, Maryland, "Shine" Parker's roommate, was the first Eagle to be shot down in 1942. He was also the first member of the second Eagle Squadron, 121, to become a POW.

He was hit by enemy fire on a sweep over France on March 8. When last seen by his squadron mates, his plane was losing altitude, trailing a long streak of glycol fumes. Jones said later he tried to ram an Me 109 as he went down, but he missed the German plane by only about four feet. He then found himself imprisoned in Stalag Luft III.

Shortly after 121 Squadron received notice that Casey Jones was alive and held at the Sagan prison camp, his friend Leroy Skinner contributed greatly to a 12-day Squadron winning streak. In three engagements in that period, the 121 pilots shot down four enemy planes and damaged four, and Skinner figured in all three fights. He shot down one German plane in the first battle, shared with Selden Edner in a FW 190 kill three days later, and on April 24 teamed with Jim Daley to shoot down a Ju 34 trainer.

Four days later Skinner himself was shot down while escorting Boston bombers striking a target at St. Omer. Another 121 Squadron member, Carl O. Bodding, tried to bail out at low altitude, and his plane was seen crashed in flames. Skinner later joined Jones and the other fallen Eagles at Stalag Luft III.

On May 10, 1942, Royce Clifford "Wilkie" Wilkinson, commanding officer of the new Hurricane fighter-bomber squadron at Manston, led his aircraft, each carrying two 500-pound bombs, against the strong Luftwaffe Me-109 fighter base at Abbeville, only 75 miles across the English Channel from Brighton, England, and about 125 miles from London.

Two years earlier, before the first Eagle Squadron was formed, Wilkinson had become an ace—in fact, almost a double ace, credited with nine confirmed kills. As a sergeant pilot with RAF No. 3 Fighter Squadron under Walter Churchill, Wilkinson shared with five other pilots in the destruction of two Nazi planes on May 12, 1940. His score for this engagement: one-third of an enemy aircraft. The following day he shot down another plane, and on May 14 brought down two Me 109s. He scored another victory May 20. By the time he returned to England, he had destroyed five more.

Leadership positions soon followed for this high-scoring fighter pilot:

> When we got back to England, Churchill was promoted to squadron leader and carried on as CO. I was commissioned and made F/Lt as officer in command of A Flight. Walter got the Distinguished Service Order, and I got the Distinguished Flying Medal and Bar.
>
> When 71 Squadron was formed September 29, 1940, with Churchill as CO, he insisted on picking his own flight commanders, so George Brown and I were posted to Church Fenton to form the Squadron. (Both Churchill and Brown were recipients of the Distinguished Flying Cross.)
>
> We were given some Brewster Buffaloes, which were originally destined for Belgium. After a few flights in them, both Walter and I decided they were useless as an operational fighter. After several rows with the Air Ministry, we were given some Hurricanes.

When the second Eagle Squadron, 121, was formed at Kirton-in-Lindsey May 14, 1941, Fighter Command transferred Wilkinson to it in the same flight leader role he had occupied in 71 Squadron. In March, 1942,

Wilkinson was promoted to the command at Manston. In May, he and his squadron cohorts found themselves descending, bomb-laden, on the Abbeville target. He recalled that day as follows:

> I came in low to machine-gun Abbeville drome, and just as I was pressing the tit to release bombs I heard and felt a hell of a crash sound. My first thought was that a stupid bloody armorer had put in the wrong fuse and I had blown myself in. But there was a built-in 11-second delay to allow time to get away.
>
> I pulled up over, up to 2,000 feet from 10 feet off the ground, to bail out. I tried to step out on the side of the spin, got back in, kicked on full left rudder, and got the plane out of the spin. I yanked it out, clipping the treetops as I pulled up, and just as the plane was going into a stall at 800 or 900 feet I nipped over the side.
>
> I saw the aircraft, all the front of it gone, just above me, and flak pouring into it. I fell through the trees. The chute, just opening up, tangled in the trees and lowered me gently to my feet. The aircraft crashed about 50 yards from me. I almost beat it down.

Wilkinson stuffed his Mae West under a hedge in a wood about half a mile away from the airdrome. Then he hid in a ditch. A truckload of German troops drove straight to the burning Hurricane and started searching through the trees.

> They started in front of the wood and I followed behind them. They searched, and I was not there. Finally they cleared the wood. I hid in a ditch by a farmhouse until dark. I saw them search the farmhouse and go away. I went into the house after dark and told the farm family who I was.
>
> They fried some eggs and gave me some other food but told me they couldn't help me more than that. They said they didn't dare because the Germans would be back to continue their search.
>
> The farmer gave me a jacket to put on, and they stuck my RAF jacket into the fireplace. I went back into the woods and found another farm. They gave me more eggs, and then I set out to walk to Abbeville railroad station.

A more cautious pilot might not have dared to approach anywhere close to an area he had attacked only a few hours earlier. For Wilkinson there appeared to be no acceptable alternative. He was still wearing RAF trousers and an RAF shirt, but the farmer's jacket helped take away the military appearance. He picked up an old sack and flung it across his husky shoulders, peasant-wise. At a height of 5'10½" and weighing 190 pounds, the brown-haired 28-year-old Yorkshireman felt that he reasonably well resembled an ordinary French workman. He had picked up some elementary French. He walked to the station and bought a rail ticket to Paris.

In the French capital he set out to find help. "I went into a bar, trusting that everyone there would be French," he said. "The barman gave me a ticket on the Metro that would get me around the city. The barman told me to come back at 6, but when I did he said he was unable to make a contact here. He advised me to go to Chagny on the demarcation line between German-occupied France and Vichy France." Chagny, in eastern France, is 75 miles north of Lyon and about 115 miles west of Bern, Switzerland.

In Paris, Wilkinson went into a barbershop. Two German officers came in and ordered him out of the chair so that they could be served.

> I replied in Arabic, and they didn't know what language I was speaking. They left me alone. The barber shaved me—he guessed who I was—and there was no charge for the shave.
>
> Looking for a place to sleep, I went to what turned out to be a brothel. The people in charge asked if I was a German. "No, I am an RAF escapee," I told them.

The next morning he bought a ticket and boarded a train to Chagny. At that stop he left the train and simply walked out a gate. German guards at the gate let him pass through without question.

> From there it was 20 kilometers to the frontier. Villagers treated me to wine and eggs. Some young kids showed me the way to the demarcation line, and told me a guard would be coming back to that point about every five minutes. I threw some sacks across the barbed wire so that I could climb over it, and was at last outside of the German-occupied part of France.
>
> A local gendarme drove me to Macon, about 45 miles south of Chagny. There I stayed in a hotel run by a French woman with a Belgian and a Frenchman. They had been in the French secret service. The woman took me under her wing.
>
> A couple of men came to the hotel to see me. After talking with me for half an hour they said, "We are the French Secret Service. We are not working for the Germans; we are working for you." They made some photographs and produced identity papers, and they said I would be using the name of Leon Boyer. Many years later my friends in France still called me by that name.
>
> The French captain said, "We don't know your escape route. What can we do?" I suggested that they get me a boat. I could take it out in the Channel and get back home that way. I went to Chamonix, and for a while was hiding out in the general post office. After a couple of weeks a message came through that they could not get a boat for me. A woman came and said she would take me to the American consul in Lyon. She was Virginia Hall, a war correspondent for one of the American papers, a strapping big girl who was in the Underground, working with the Resistance.
>
> I stayed for a while in the Lyon flat of the U.S. Consul, Ron Whittingham, or Whittaker. Just after that the Germans occupied Marseille. I got in touch with the French escape system through General Pat O'Leary Gerard, a Belgian doctor, and got out over the Pyrenees and on to the British embassy in Barcelona about five weeks after bailing out. I got together with two Poles and a Frenchman from North Africa, and we agreed we had to get to Madrid. From outside the door of the Embassy in

Barcelona, we stole a Spanish Ford V-8, with full fuel tank and a couple of Jerry guns, and drove all through the night, arriving in Madrid at 6 A.M. Had we arrived a bit later we would have had trouble, because the Spanish guards came on duty at 8:30. I stayed there in Madrid a few days, changing my rank to private in order to avoid the Gestapo. I was repatriated to Gibraltar and hitch-hiked home from there.

Wilkinson reported back to England in July 1942 for reassignment. Initially the RAF proposed that he be given a noncombat posting; if the Germans should shoot him down again, he probably would be executed. Wilkinson made it clear that such an arrangement would not be to his liking. He was sent to 56 Squadron, flying Typhoon fighters. In August he was promoted to command another Typhoon squadron, No. 1, at Biggin Hill.

Over the years Wilkinson never forgot the heroism and sacrifice of the French people who helped him, sheltered him, and moved him along escape routings in a terrible time. "The French were starving, yet they fed me," he recalled. "They laugh a lot, sometimes with a distorted sense of humor, because they believe that laughing makes you fat. I love the French people, but not necessarily did I love their government of that period."

* * * *

Robert Lee Priser, who came to 71 Eagle Squadron at Debden in the spring of 1942, recalled that at that time all the pilots were concerned about the new German fighter, the Focke-Wulf (FW) 190. "It was proving difficult to contend with in combat," he said. "Sam Mauriello took us aside and emphatically told us the rudiments of air combat, such as 'always keep a sharp look out for the enemy,' 'always turn into the attack,' 'never reverse a turn with the enemy on your tail,' and 'you can't outrun a bullet but you can sure as hell out-turn it.'"

Oscar Coen, the first successful Eagle evader, had lost little time getting back into combat. On April 27, 1942, he and his best friend in the squadron, Michael G. "Wee Mac" McPharlin, shared in shooting down three of the FW 190s. Each man good-naturedly grumbled that he was credited with only half a kill per plane, instead of the full scores he would have received if his partner had only stayed out of the way. Coen and McPharlin thought alike, fought alike, and were about the same size. They agreed that if either was shot down, the other would take his pick of the loser's gear. "Mike had a pair of boots I liked, and he wanted my leather suitcase," Coen said.

Four months later, covering the Operation Jubilee commando landings on either side of the French coastal city of Dieppe on August 19, Coen shot down another FW 190, and he and McPharlin damaged and probably destroyed a Junkers 88 bomber. This time McPharlin's plane was hit and he had to bail out, as did Squadron Leader Peterson, who had just shot down a Ju 88. Both men, along with Gene Fetrow, were quickly rescued from the Channel by British naval vessels.

A fourth Eagle, Julian Osborne, also bailed out but came down in

England. In the confusion of fiery combat, James Taylor and another Spitfire pilot accidentally fired on each other, and both were killed.

When Peterson and McPharlin arrived safely back at the base, the latter marched up to Coen, who was trying on the spoils of his buddy's misfortune, and demanded, "Oscar, take off those damn boots!" (Their friendly-spirited rivalry was to continue nearly two more years, until June 6, 1944—D-Day in Normandy—when Mike McPharlin was shot down and killed.)

Although Coen and McPharlin were battle-hardened to the point they could make light of McPharlin's downing, Priser, who was engaged in only his second operational flight since reporting to 71 Squadron at Debden in June, recalled the Dieppe raid in awe.

"I flew tail-end Charlie to a flight of four Spitfires, and the first thing I saw over the area at 12,000 feet was four parachutes in the air and five or six dye marker patterns in the water where other parachutists had come down. We stayed over the area for almost 40 minutes, the air filled with aircraft from both sides. Below, I could see the German guns on the shore pounding hell out of the withdrawing Allied barges. I don't know how any of them managed to get back to England."

One of the parachutes belonged to Jackson Barrett "Barry" Mahon, the six-plane ace from 121 Eagle Squadron. In his eight months of fighter-pilot service with the Royal Air Force, of which only the latter half involved combat, Barry Mahon had become a hero and something of a legend in his home town of Santa Barbara, California. He had damaged his first enemy aircraft, an FW-190, on April 12, 1942, and within six weeks had poured damaging fire into three other German fighters, and had shared in the sinking of an armed trawler and a 2,000-ton mine sweeper.

In a sharp engagement June 2, he and squadron mate Jack Mooney had each shot down two 190s. On July 31, with 121 Squadron marking up its highest score—seven enemy fighters destroyed—Mahon and Sel Edner had each shot down two planes, and Squadron Leader Hugh Kennard, William P. Kelly, and Frank R. Boyles had accounted for one more each. Norman Young, a pilot new to 121 Squadron, had been killed. On a separate bomber escort mission that same day, 133 Eagle Squadron had accounted for three enemy planes, but three of its pilots—Carter Harp of Georgia, Coburn King of California and Grant Eichar of Illinois—had been killed. Edwin D. "Jessie" Taylor had been gravely injured.

Now, floating in the Channel in his dinghy, watching 80 or 90 German planes and an equal number of British aircraft doing battle in an area five miles across—"probably the greatest air show the world had yet seen"—Barry Mahon gradually drifted ashore into the hands of the Germans. His captors took him to a prison camp near Paris and turned him over to the Luftwaffe.

> As their first American prisoner, I created a sensation. One pilot after another tried to buy me drinks, trade wings, or just practice English on me. An intelligence officer tried to question me, but the fliers refused to let me go.

In the evening some of the Luftwaffe pilots took Mahon into Paris and left him in a truck under guard while they toured the cabarets.

> They said they were afraid of the SS officers in Paris, but they kept coming back to give me glasses of champagne. I was beginning to wonder why we fought the Germans at all. They all seemed like very nice fellows—until they turned me over to the SS, the Gestapo troops—an entirely different type of men—for interrogation. They were sadistic, mean and arrogant.

Transferred to an intelligence center just outside Frankfurt, Mahon found himself in a structure called "the cooler," housing cells for the solitary confinement of prisoners.

> They presented me with an innocent-looking Red Cross form—name, rank, serial number, and so forth. In the second section it went into a little more detail. What type of aircraft, number of guns, did it carry bombs, number of people in the squadron, losses in the last 30 days. I refused to fill out the last section and was told that until I did so I would remain in solitary confinement.
>
> I stayed in the cooler for three weeks, and then they came and said they were sending me to another camp. I am sure that even if I had filled out the questionnaire, my information would not have been too enlightening. During the three weeks, they showed me scrapbooks about the Eagle squadrons they had compiled in their intelligence division, including radio-telephone language, which had been recorded daily. I must say it does not make for any great feeling of security when you are in the middle of Germany to be confronted with a statement such as "I just shot up two of the dirty sons-of-bitches." The Germans seemed to take that type of thing as a matter of course, however. I do not believe I was treated any the worse for it.

From the interrogation center, Auswertestelle West, Mahon, like other pilot POWs, was sent to Dulag Luft, the distribution center in Frankfurt proper, and then on to the Stalag Luft III permanent prison camp.

From his unique vantage point, Bill Hall also took notice of the impacts of Operation Jubilee:

> In July 1942 I was moved from Hildburghausen to Oflag IXA/H, a former Hitler Yungen Camp at Kassel, Germany. There were two camps at the town: IXA/H in the center of Kassel by the river, and IXA, known as Spangenberg, up on the pinnacle. It was a medieval castle with a high stone wall, a moat, and a drawbridge.
>
> At IXA/H there were two barrack blocks with an 'abort'—a toilet building—of 20 stalls, back to back, separated by a low partition. In our building we had a mess hall and kitchen downstairs. Upstairs there were two or three large rooms with double bunks, wooden slats, straw pallets and one blanket each, with about 40 men to a room. The only window in our room looked out over the parade ground, and at the end of the barbed

wire fence was one of the raised guard houses, about 15 feet off the ground, armed with machine guns and rifles. To the right of the window were the fences, the river, and the road leading to the main square of the town, with houses on the other side of the road.

Every morning the German women would hang their bed clothes over the balcony railings to air. About the time I arrived at IXA/H the prisoners from Dieppe started coming in. They were Aussies, New Zealanders, and Canadians, which really filled the camp.

At this time word came through to the German High Command that some German POWs had been mistreated, so the Germans pulled a reprisal on us. They removed all our shaving gear and tied our hands with rope. This lasted about a week. When we had all grown beards, the Germans came in and took pictures of us. Then, after the shaving gear was returned and we shaved, they moved in again for new pictures.

One bright and sunny morning, Cappy Blair Hughes-Stanton, an English army captain, and I were watching a German girl hanging out her bedding. I whistled at her and at that moment saw the sun glint off a rifle barrel to our left. I hit the deck and shouted to Cappy. Cappy was a little slow and was shot through the mouth, losing two molars. The Germans were fairly trigger-happy then, the guards being wounded men back from the Russian front. Anyhow, Cappy got over it and was left with only a small triangular scar on each cheek.

At this camp we were given a diet of salted fish. I understood it had been caught in the North Sea and it was brought in by the truckload. It looked and tasted like cordwood and had to be soaked for three days before we could eat it. But then the Germans messed up and brought us a whole truckload of canned cheese from a cheese factory that had been bombed. The cheese was all canned in two-quart cans that had been through fire, and it was the nicest cheese that anyone ever tasted, mild and yellow. After we had eaten quite a bit of it, the Germans found out how good it was, and they came in and picked up all that was left.

So our diet was mainly the fish, three boiled potatoes, a little bit of pure white margarine that looked like axle grease but didn't taste too badly once you got used to it, and a tenth of a loaf of black bread. Sometimes they would come around with what they called meat soup, which was made with horsemeat.

When the Red Cross parcels came in, our menus improved. Sometimes we'd get one a week. As more POWs arrived, the parcels were spaced out.

Bill Geiger remembered his time in the old castle in Kassel as "my first experience with a great deal of hunger." As he recounted:

Most of the prisoners there—just a few hundred—were on the staff of General Fortune, commander of the British forces in France at the time of the fall of the French government. They were all older, and they didn't eat too much. They had been receiving parcels from friends in America for quite some time and had quite a store of them. The Germans had gotten used to backing off on the amount of food they were giving them because they really didn't need it.

So you put 60 very hungry young Air Force people in there, and we just didn't have enough to eat. There wasn't really any way Fortune's people could feed us. It was just pretty miserable. We were stomach-ache hungry for five or six weeks.

Hunger and discomfort could not stifle the antics of the practical jokesters at camp, like Bill Hall, who happily brought indignity to the gentlemen British officers in camp.

Before the Aussies arrived at IXA/H it had been a senior British Officers' camp, mainly prisoners from the Dunkirk evacuation. Of course, the prisoners had everything established in the British fashion. At IXA/H we had roll call every morning at 9, and again just before dark. In the morning the British officers—brigadiers, colonels and other senior ranks—would line up in front of the aborts and as soon as roll call was over they would occupy all the spaces in the abort, taking all the time in the world while we younger fellows had to wait our turn.

The building seated about 40, 20 on each side of the partition, back to back, with a creek underneath providing a supply of running water. The situation of the senior officers taking over the aborts prevailed every morning for some time until a couple of us younger fellows played sick during roll call and sat at the top of the creek in the abort building, out of the line of sight. We each had a good bundle of paper and excelsior, so when roll call was over and the officers had moved in and were comfortably seated, we lit the excelsior, dropped it on the water and let it float down stream. As it singed each bare bottom, the senior officers leaped off the seat and out, with their pants around their knees. From then on, there were always vacancies in the abort after roll call.

In all, IXA/H was a "good POW camp," from Hall's point of view. He was to remain at Kassel until January 6, 1943, when he was moved to Stalag Luft III. When he reached there, he would find many more Eagles caged within its fences.

8

The Disaster of Morlaix

ON SEPTEMBER 29, 1942, the Eagles officially transferred from the Royal Air Force to the U.S. Army Air Force. As Oscar Coen recalled, the event ushered in a sense of dual pride in their service to England and now to their home country.

> Since American equipment was not immediately available, the Eagles retained their Spitfires for a transition period and painted U.S. Army Air Force stars over the RAF red, white, and blue roundels so that the Germans and everyone else might take note of the change-over. The pilots changed into American uniforms as soon as they had time to acquire new clothing. They were proud to attach the silver wings of the USAAF to the left breast pocket of their jackets, and they felt honored when the USAAF granted their request to be permitted, as former members of a British unit, to wear miniature RAF wings over the right breast pocket.
> The British presented us, by way of the Duchess of Kent, a silver medallion to be worn on the left pocket under our new silver wings. They were cast only in a specific number, and we were proud of the casting. We had permission to wear them also as a part of our U.S. uniform.

Yet, as Frank L. Kluckholm reported in the *New York Times*, other emotions played out at the change-of-command ceremony on September 29:

> Tragedy tinged the scene as members of three Royal Air Force Eagle Squadrons—the first American fliers to get into action against Adolf Hitler in Blitz days—entered the Air Forces of their own country with impressive ceremonies today.

For only a brief time before changing from the Royal Air Force blue to the United States Army olive drab—on Saturday—some of their number were lost in action over France.

It is now possible to disclose that some Eagles were among the RAF men lost in 11 planes because of bad weather that caused wing icing.

The German radio asserted that four Eagles were captured after the planes came down, listing Lieutenants Charles Albert Cook of Alhambra, California; Marion E. Jackson, Corpus Christi, Texas, wounded; Edward Gordon Brettell, England, wounded; and George B. Sperry, Alameda, California, wounded.

The Eagles were grim and solemn, therefore, as they stood by at salute in the square surrounded by camouflaged buildings and wearing RAF uniforms for the last time as the Stars and Stripes fluttered up the pole to wave next to the RAF flag while the band played "The Star-Spangled Banner." Their faces showed they had lost some of the joy that the day had promised. . . .

In fact, the ill-fated events described so facilely as weather-related were to lead not only to the imprisonment of the four Eagles first announced by the Nazis, but to a wild flight through enemy territory for three others, with varying results—including, for one fugitive, an unexpected reunion with another downed Eagle.

On their mission of September 26, 1942, the fighter pilots were assigned to safeguard American-manned B-17s bombing railroad yards and an aircraft maintenance plant at Morlaix, on the Channel coast of France. Submarine pens at Brest, 30 miles away, were a secondary target.

Unable to find their targets in adverse weather, the B-17s jettisoned their bombs and flew home. After bidding goodbye to the bombers, the Spitfire escort went astray over a solid cloud blanket extending as high as 6,000 or 7,000 feet above the ground. Through a break in the clouds a shoreline became visible; supposedly, it was the coast of England. When a city came in view, Flight Lieutenant Gordon Brettell, leading the squadron, assumed that it was English. Accordingly, the Spitfires, all low on fuel, passed in an orderly close formation over the town, at about 2,500 feet. It turned out to be not Southampton or Plymouth but Brest—the site of German wall-to-wall antiaircraft guns and fighter bases. In a moment the sky was full of bursts of flak—large, medium, and small. Immediately, that precision formation exploded into individuals flying for their lives, all air discipline forgotten.

Earlier, the Spitfire piloted by Gene Neville had developed engine trouble, and another pilot, Dick Beaty, had been ordered to escort him back to England. Neville was shot down and killed 40 miles east of Morlaix. Beaty managed to get back to England but crash-landed, out of fuel.

All together, the explosions over Brest took a frightful toll: all twelve of the newly delivered Spitfire IXs of 133 Eagle Squadron were lost; four pilots were killed; and six became caged at Stalag Luft III.

* * * *

Charles Cook—"Cookie" to his fellow pilots of 133 Squadron—was a tall, good-looking Californian who used to play football at Alhambra High School and California Polytechnic at San Luis Obispo and spend summers as a lifeguard at the Alhambra city park plunge. As an Eagle, he was a popular and well-tested veteran.

When Cook failed to return from the ill-fated Morlaix mission of September 26, 1942, he left behind in his quarters a full, detailed record of his military flying experience. It was contained in RAF Form 414, a blue hard-cover Pilot's Flying Log Book. Months afterward, on a visit to the United States, the man who had been Cook's squadron commander, Carroll W. McColpin, personally delivered the log book and Charles' other effects to the missing flier's family in California.

Cook's record book, neat and complete, carefully hand printed, legible down to the last item, listed his flight training time in the United States in 1940 and 1941 before going to England, and his 53½ hours of advanced instruction at the RAF's 56 Officers' Training Unit in Lancashire.

One page certified that on October 8, 1941, Pilot Officer Cook completed the training with grades of average in air gunnery and above average in ability as a fighter pilot. Another showed his posting October 18, 1941, to 133 (Eagle) Squadron at Eglinton in North Ireland. On yet another page Cook certified that "I understand the controls and cockpit drill of the Hurricane Mk II as explained to me by P/O Meierhoff." A month later he certified that he had received Spitfire II cockpit instruction, "and I fully understand the petrol, oil, coolant and ignition systems."

On August 21, 1942, the pilots of 133 Squadron exchanged their Spitfire Mark V aircraft for the powerful new Spit IX. They were the first Eagles to be assigned—and, as it turned out, the only Eagles to fly—the advanced Spitfire.

"When we got those Spit IXs we finally thought we were hot pilots," Cook said later. "That plane had so much power you could turn its nose up from zero degrees and it would hang on the clock (prop) and climb to 25,000 feet. I never thought I was much of a pilot until the day I pointed it up like that and saw it just hang there, climbing."

The last entries by Cook in his flying log were in September, 1942:

> Sept. 4—P/O Miley and I airborne at 5:45 A.M. to patrol from North Forland to Dover, 500 to 1,000 feet. Visibility nil, and then our own ground forces almost shot us down. Very intense flak. Two 190s bombed Hythe.
> Sept. 8—Rendezvous with 12 USAAC First Fighter Group P38 Lightnings at 25,000 feet over Tangmere. Squadron of P38s on their first look at the French coast. The P-38s were 15 minutes late for r/v, so we did not get too close to the French coast. They could stay up longer than our Spitfires could.
> Sept. 23—Squadron moved from Biggin Hill to Great Sampford, Debden, to change from 133 (Eagle) Sqdn. to USAAC 336 Sqdn.

Cook started an entry September 26: "Spitfire 9B BR640." Someone else filled in the last line:

> In preparation for B-17s' show to Morlaix, Cookie landed in France and is reported prisoner by German radio.

At the bottom of that page Squadron Leader McColpin, never knowing whether he would ever set eyes on his friend again, wrote:

> Tough luck, Cookie. It's one helluva way to lose such a good pal, but God willing I'll be seeing you after this mess is over. 'Bye now. C. W. McColpin.

It had been Cook's practice, in his methodical way, to painstakingly prepare in red ink, shortly before the end of each month, a summary of that month's operations, with the numbers to be filled in after all the statistics had been gathered. At the end of September 1942, four days after Cookie's disappearance, his friends in what was now USAAF 336 Squadron filled in the cumulative figures for him.

They showed 243½ hours of Spitfire time, 58 hours of Hurricane time, 3 hours of Spitfire night time, 125½ hours of operational time, and 99 operational sorties. One more mission would have made the total for Charles Cook an even 100.

In the German camps each war prisoner was handed a package containing among other articles a bulky album, its pages blank, identified on the cover as "A War-Time Log," a gift from the World's Alliance of Young Men's Christian Associations, Geneva.

Page One of Cook's logbook, topped with white stars on a blue background, had a black border with this inscription:

> In Memorium
> 1st Lt. Bill Baker, Cameron, Tex.
> 2nd Lt. D. D. Smith, Redding, Cal.
> 2nd Lt. Gene Neville, Oklahoma City, Okla.
> 2nd Lt. L. T. Ryerson, Boston, Mass.
> Killed near Brest, France, Sept. 26, 1942, due to unforgivable circumstances while flying Spit 9's.

Cook opened his logbook diary thusly:

> I was shot down at 6:15 P.M. Saturday evening Sept. 26, 1942, while trying to crash-land in a meadow as my petrol was just about all used up. As I was on my final approach to the meadow, an FW 190 shot my right wing off from a distance of 50 yards. I was just lucky enough to get out of the Spitfire cockpit, open my parachute, swing twice, and land in a French country hedge near Brest, France. Brest-Guipavas.

The FW pilot was a young Oberfeldwebel (German sergeant pilot) and I was his third victory. He belonged to the No. 7 Richthofen Squadron who was commanded by Oberleutnant Stolle (a first lieutenant in the German Air Force). He later became a Hauptmann (Captain) with the Knight's Cross of the Iron Cross.

Hauptmann Stolle and his squadron treated me very well. I had a grand time living and talking with them for three days before I was sent from Brest to Paris and then on to Dulag Luft at Frankfurt, Germany. I arrived at Stalag Luft 3 Oct. 17, 1942.

Elaborating on his diary 40 years later, Cook recalled that a German patrol with dogs took him into custody within half an hour of his parachute landing. The patrol took him to the local German Army headquarters, from which the Luftwaffe removed him to No. 7 Richthofen Squadron.

Bruno Stolle, the officer in charge, told me I looked like one of his own pilots. I told him my ancestors were German; the name had been Koch. There were newspapermen and cameras around.

They let me look at an FW 190. Their pilot flight room was like ours; the operations board was exactly like ours. It was amazing how much our squadrons resembled each other. But they had better clothes. They had on leather pants, boots with flares and guns, jackets—beautiful stuff, and everything neatly pressed.

The next morning, here came Sperry in a Mercedes. Soon they brought in Jackson, on a stretcher. He had shot down an FW, and then the German pilot's buddy shot M.E. down. He had shrapnel in his face and hands and was very ill. The German doctor asked whether he had had a tetanus shot. When I said he had, the doctor said, "Good—I'll give him another." They took him to Paris, where he met Brettell. They were in the hospital in Paris for about six months before being sent to Stalag Luft III in Germany.

Meanwhile a big truck drove up with the bodies of Ryerson and D. D. Smith. Neville's body was already there, in a morgue. They buried those three men, and the German pilot that Jackson had shot down, all four of them together in the same cemetery.

* * * *

George Sperry's arrival in a Mercedes was the result of an action of an informant. As Sperry recalled the Morlaix mission and its aftermath:

I was at 7,000 feet, just a little east of Brest harbor, with the canopy of my Spitfire shot away, when the overheated Merlin engine seized and quit cold. I took my helmet off, tossed it on the floor, opened the side door of the Spit IX, stepped out on the wing root, and reached in to push forward on the spade grip, as the airplane was close to a stall. I shoved myself away from the fuselage, and as the tail went by I pulled the ripcord of my chute. I landed several minutes later in a French cow pasture.

Although the ground was soft and muddy from recent rains, my contact with the surface was far from gentle—a real bonecrusher, due to several factors. The day before, I had taken my own chute, which had a

28-foot canopy, into the parachute loft to be repacked. A substitute 24-foot canopy was fitted to my 196-pound frame. With no indication I would be using it within the next 24 hours, I gave it no further thought.

The 30-knot wind blowing across the pasture was another matter, as I came down very fast, drifting with the wind. I hit the harness quick-release as my feet touched the ground. As I went into the mud I became entangled in the shroud lines, and the chute dragged me through the mud. I ended up next to a six-foot hedge row in the middle of a very large pond of water. Struggling to my feet, I gathered in the fabric of the canopy and the shroud lines and stuffed the whole bit into the hedge row.

I came around a break in the hedge and there stood a French farmer. He asked me who I was. Remembering some schoolboy French, I replied that I was an American flier. He motioned me to follow him, and we walked about a quarter of a mile to a farmhouse where I could wash the mud and blood from my face and quench my thirst. Evidently when I lost the Spitfire canopy a small piece of something had given me a cut high up on my forehead.

Some minutes later another Frenchman came in the door and to my surprise gave me the word, in excellent English, that in order not be captured by the Germans I must start out immediately, moving in a southeasterly direction. In the next few hours I managed to cover quite a few miles, running 50 feet or so and walking 50. I had figured rightly that the French Underground would not touch me with a six-foot pole unless I could get far inland from the coastline.

After dark I stayed close to the road. Once, when I heard a vehicle a short distance away, I crossed a fence and hid on the ground behind a tree. Later I could hear dogs in the distance, but I did not get too worried because a light drizzle was falling and probably would cover my tracks. It was the motor traffic that scared me.

Around 2300 hours, drenched to the skin and starting to feel a little sorry for myself, I stopped alongside a tree for a short break. Something touched my arm and a very calm voice whispered in my ear, "It looks like you've had it. One of my countrymen has informed the Boche of your whereabouts."

"I would like to get my hands on that son of a bitch," I said. Coolly he replied, "The informer will be taken care of. Now you had better get on down the road abut 50 meters and into the farmhouse you'll see on the right. I am sorry you could not have gone far enough, in order that we could hide you safely. Good luck to you."

He slipped into the darkness and I walked down to the farmhouse, went in, sat down next to an open fire, and lit a cigarette. A few minutes later, German soldiers had surrounded the place. I was still soaked completely through, and chilled to the bone. The Germans escorted me to the local French jail, and I spent the rest of the night pondering the situation, wondering what had happened to my buddies. I was afraid that Gordon Brettell had "bought the farm," and maybe Bill Baker had, too. Several hours earlier I had heard Bill on the r/t—the radio telephone—say he was flying up through the overcast, and then calling out, "I'm out of gas and going to ditch in the drink." Bringing a Spit down in water was ill advised and almost always fatal (as it had been for Baker).

Early the next morning, September 27, a Luftwaffe captain took me out of the jail and escorted me across the street to an army headquarters. Here I had a shave and a hot bath, a short nap, and then a fine breakfast. During this time the Germans had washed my clothes and cleaned my muddy uniform jacket and trousers. Then I climbed into an open staff car with the same Luftwaffe officer and was driven to the Brest airfield. As we rolled up to the pilots' mess there I saw, standing on the front steps and surrounded by the pilots of the 7th Richthofen squadron, Charles Cook, Jr. Cookie quickly gave me the sorry news: Gene Neville, Leonard Ryerson, and Dennis David Smith had been shot down by antiaircraft batteries and were dead; M. E. Jackson had been shot down by the same flak battery, but before crashing in a vineyard alongside the airdrome had shot down and killed an FW 190 pilot as he was taking off. Now Jackson was in a station hospital with flak wounds.

Sperry also learned in some detail of the very narrow escape of Cook himself:

Out of fuel, he had jettisoned his canopy and was prepared to belly his aircraft into a field in front of him. At this moment an FW 190 closed in from behind and started firing, and the right wing of Cook's plane flew off. Cook pulled the pin on his Sutton harness, pushed forward on the spade grip, fell out of the cockpit and opened his chute. Seconds later he touched down right in the middle of an antiaircraft battery.

Sperry, too, remembered the funeral of September 27, but explained that his grief had to be held private:

About noon on September 27, the day after our mission disaster, Neville, Ryerson, and D. D. Smith, and the German pilot shot down by Jackson, were buried with full military honors at the local cemetery. Cook and I were not allowed to attend the funeral for our friends, the explanation from the Luftwaffe being that our presence might cause disturbances among the French population.

But the day had its share of frivolity as well, as Sperry was quick to point out:

That same afternoon Oberleutnant Bruno Stolle, 7th Staffel squadron commander, showed Cook and me over the airdrome. After inspecting the cockpit of his FW 190, I asked permission to fly the plane around the field a few times. This request was very emphatically denied. That evening, in the best fighter pilot tradition, the party thrown for Cook and me at the pilots' mess turned out to be a huge success, much like our nightly brawls back at Biggin Hill. We were gently lowered into our sacks, full of brandy and champagne, some time during the night.

Late in the afternoon of September 28, Cook and Sperry were driven to the Brest railway station to begin their trip to Germany and a prison camp for the duration of the war. Sperry found their departure "sort of like going on leave except for the three-man Luftwaffe guard detail. The squadron pilots came to see us off with presents of wine, fruit, sandwiches, and cigars so that we would not want for anything on the journey."

The next morning, the fallen Eagles arrived in a suburb of Paris, very much aware that this was the day that 71, 121, and 133 Eagle Squadrons were officially becoming 334, 335, and 336 squadrons of the Fourth Fighter Group. George Sperry recalled his thoughts: "I remember the date—September 29, 1942—very well. After a short breakfast, escorted by our guards, we wandered about the town throughout the rest of the day. Some Frenchmen occasionally slipped us cigarettes, and others as we passed them gave the V for victory sign. We were both aware that the ceremony, the change-over that same afternoon at Debden, was going on, attended by Air Chief Marshal Sir Sholto Douglas and Major General Carl "Tooey" Spaatz, and that it would see the Union Jack hauled down and replaced by the Stars and Stripes.

"That night on the train going toward Germany a thousand thoughts ran through my head. I guessed I was lucky just to be alive, and now this was the final stroke of fortune, to be spared to become a prisoner of war instead of a charred corpse. What a hell of an ending for a trained fighter pilot, unable to be of real use to my own country's air force. I had to learn to accept the situation and to make the best of captivity, to learn the meaning of patience."

On the morning of September 30, Cook and Sperry arrived at Dulag Luft, the center for Luftwaffe interrogation near Frankfurt on the Rhine. The German intelligence officers had complete files on all Allied flying personnel. Here, during solitary confinement for the next six days, Sperry was enlightened as well as interrogated in a small 6 x 12 room containing two stools, a small table, and a bed. The window was locked and the air was very close.

> The first interrogation consisted of giving the intelligence officer my name, rank, and serial number. Before leaving he closed the window and locked it also. Within the next four hours or so the room temperature went higher and higher. Three or four hours later the same officer came back to interrogate me and found me sitting on the stool stark naked, wet with sweat and mad as hell. Interrogated again, I gave him my name, rank, and serial number and added a few choice words on my estimate of the present situation. So I spent another 24 hours getting the heat treatment.
>
> The third day guards unlocked the door and escorted me to the shower room, where I managed to cool my hot temper somewhat. Dressed again, I was returned to my hot-box cell to find the window open and the intelligence officer seated on one of the stools. As I sat down he offered me a Camel. From under his arm he took a large book labeled Adler Staffle—Eagle Squadron—, opened it to my name, and showed the personnel file for George B. Sperry. Talk about being enlightned. I was astounded.

It showed where I was born, facts about my family, where I was educated, where I learned how to fly in 1935, and all about my enlisting and training for the RAF. My letters home for the last year and 10 months, which my mother had sent to the local hometown papers, apparently had made it very easy for German intelligence officers to follow my career as a fighter pilot. When the intelligence officer left my cell, I never saw him again, but I did spend three more days sweating it out, probably as punishment for my uninhibited language on my first interrogation. Released from solitary, I was sent to the Compound, where I received clean underwear and socks and was allowed the privilege of a shave and shower. Cook had preceded me by two days.

* * * *

George Middleton found out quickly that the worst part of his Morlaix mission was his parachute jump. "Supposedly, when bailing out of an airplane, one grasps the ripcord and counts slowly to ten before pulling it," said Middleton of his departure from his new Spit IX over France, with the fuel gauge registering zero at the unplanned end of the disastrous final mission of 133 Eagle Squadron September 26, 1942.

I was only about a thousand feet above the ground, so my procedure was "one, and two, and, hell with this, pull."

About three feet of the cord came out, with six inches of bare wire at the end. I was horrified; thought I'd broken the bloody thing. I threw the ring away and spread my legs wide, prepared to tear the chute apart manually. Suddenly—pow! It opened.

One of the few things I remembered from boring lectures on parachute techniques was the warning to keep the legs together. Now I found out why. The slowing force of the parachute almost pulled me in two.

Swinging gently under the parachute, Middleton watched his plane spiral downward and explode upon the countryside, half a mile away. He saw people running toward the scene of the crash.

Some were carrying mattocks or other brutal-looking tools. I was glad they hadn't noticed me. German and British flying uniforms were discouragingly similar.

As I descended, I realized the wind was blowing me along at 20 to 30 miles an hour. I could see that the first few steps I would take after landing would be mighty long ones. Reaching back into my chuting lore, I recalled that a pull on the up-wind risers would cause the chute to slip into the wind, neutralizing the drift across the ground.

Confidently I grabbed a double handful of the up-wind lines and gave a great haul. The results were spectacular. The chute must have almost collapsed, because I dropped like a stone for about 200 feet. The chap who wrote that part of the manual either never had tried it or had left out something important.

The closer I got to the ground, the more it appeared that the first giant step I would take would be backwards. The training manual had explained how to cope with that problem, too. Cross your arms, grab the risers, and

"Red" Tobin on the cover of the May 1941 *MGM News*.

Bill Nichols, 71 Squadron.

Nat Maranz, 71 Squadron.

P/O W. (Bill) Hall, 71 Squadron.

BUCKINGHAM PALACE

    The Queen and I offer you our heartfelt sympathy in your great sorrow.

    We pray that the Empire's gratitude for a life so nobly given in its service may bring you some measure of consolation.

George R.I.

I. A. Tobin, Es.,

His Majesty's official message of consolation to the father of "Red" Tobin.

Morris "Jack" Fessler, 71 Squadron. His Spitfire was downed in France in October 1941 (below).

Oscar Coen, alumnus of 71 Squadron, as a U.S. Army Captain.

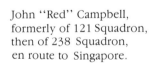

John "Red" Campbell, formerly of 121 Squadron, then of 238 Squadron, en route to Singapore.

Seldon Edner, 121 Squadron.

Distinguished Flying Cross recipients F/LT Jimmy Daley, Amarillo, Texas (3rd 121 Squadron leader); S/Ldr Hugh Kennard, Kent, England (2nd 121 Squadron leader); and P/O Barry Mahon, California (121 Squadron); in front of a Spitfire V.B.

Charles Cook, formerly of 133 Squadron.

Brewster Morgan standing beside his P-47 Thunderbolt, Debden, 1943.

Five of the POWs in this photograph, taken at Stalag Luft III circa 1943, were Eagles. (Standing, L to R) Robert Ingraham, Sam Meyers, Marion Jackson (133 Squadron), Barry Mahon (121), Roy Skinner (121), Edward Tovrea. (Kneeling) Charles Cook (133), George Middleton (133), Albert Clark, Clermont Wheeler.

Eagle Squadron pilots at Stalag Luft III in 1944: (L to R) Seldon Edner, Bill Geiger, Aubrey Stanhope, Brewster Morgan, Raymond "Bud" Care, Bill Hall, Gil Wright, Vasseure Wynn.

This monument stands outside the former POW camp at Sagan. The inscription reads: "Erected by the prisoners of war of the Allied Air Forces in honored memory of their comrades who died at Stalag Luft 3 during the war, A.D. 1939-194 per ardua ad astra." Due to the uncertainty of the length of the war, the date of the inscription was not finished.

Bill Geiger, Aubrey Stanhope, Seldon Edner, Brewster Morgan, Bud Care, and Gil Wright at Stalag Luft III in 1944.

Hank Ayres, 133 Squadron.

Kenneth G. Smith, 121 Squadron.

George Carpenter, 121 Squadron.

Robert G. Patterson, 121 Squadron.

Donald H. Ross, 121 Squadron.

R. L. "Dixie" Alexander's identification card, courtesy of the Third Reich. Alexander was one of the 133 Squadron Eagles interned at Stalag Luft III.

Former Eagles Fonzo Donald "Snuffy" Smith (left) and Jim Goodson (right) were reunited with former POW interrogator Hanns Scharff in Los Angeles in October 1983.

P/O James A. Gray, 71 Eagle Squadron, Debden, May 1942.

Frank Fink, 121 Squadron, at the cockpit of his Spitfire.

Fred Scuddary and Bill Edwards, 133 Squadron.

P/O John Slater and P/O Roy Evans, 121 Squadron.

P/O Steve Pisanos, 71 Squadron, Debden, summer of 1942.

Donald K. Willis, 121 Squadron.

Willis's P-38 after his forced landing in Southern Holland on April 10, 1944.

Captain Marion E. Jackson and friend in London, England, on V-E Day, 1945.

Jackson and Robert E. Smith, both of whom were shot down near Brest on September 26, 1942, attended an Eagles reunion in San Diego in May 1990.

The author's widow, Tess Haugland, next to the Eagle Squadron memorial in Grosvenor Square, London. The memorial, which displays rolls of each of the three Eagle Squadrons, was dedicated on May 12, 1986.

Memorial to the POWs who escaped Stalag Luft III in March 1944, but were recaptured and executed by order of Hitler. Among them was F/L Gordon Brettell, 133 Squadron, who led the ill-fated Morlaix mission, September 1942.

twist around to face the right direction. After my last experience with the cords, I was reluctant to touch them. I decided it did not really matter which direction that first step took. It would be a bitch anyway.

The ground was coming up fast. I seemed to be heading for a beautiful, soft-looking field, if only I could get over the mossy looking sod hedge in between. I knew about these spongy-looking fences. I had seen a friend make a forced landing through one of them. He wound up still sitting strapped in his seat, watching the engine roll off in one direction and the rest of the plane in the other. Farmers apparently took all the stones out of their fields and stacked them along the borders, and let the grass grow over them. They just *looked* soft!

As I neared the hedge, moving backwards, I raised my legs to clear the fence. Before I could lower them, I had landed on my rear and bounced about 20 feet. The British chute had a quick-release lock—a three-inch disk holding the ends of all the straps. It was necessary only to turn it a quarter of the way and strike it to release everything. I turned the lock and swung at it. I hit myself in the face twice and in the groin once before I connected with the disk and shed the chute. Then I chased the parachute down and hid it in weeds along a hedge that bordered a road.

Peasants were running down the road toward the wrecked airplane. I crept into the brush and watched them. After the traffic thinned, a boy of about 15 came trotting along. I gave him my best super-spy "Psssst!" He stopped but couldn't see me.

I called him over, told him I was the American pilot of the crashed plane, and asked him to help me hide from the Germans. In my supercharged state I was spouting French—words I had not even thought of since I was in high school.

When the lad recovered from the initial surprise, he clambered through the hedge to my side. He whispered something about following him and took off in a crushing lope along the hidden side of the hedge. I was young and in fairly good shape, but after 15 or 20 minutes I had to stop to catch my breath. We rested, and then went on for a few more minutes to a farm. I understood him to say he lived there with his grandfather.

We went into a barn, and the boy told me to wait. I collapsed into a wheelbarrow and speculated about how many 25-year-old white Anglo-Saxon male smokers had had heart attacks. The kid came back with a gnomelike little old fellow wearing a tam-o'-shanter and carrying a big jug and a couple of bowls. The old guy marched up to me, removed his tam and bowed, and tugged his forelock in the process. I had read about that but thought it had died out with the *droit de seigneur*.

All I could think to do in return was salute. It must have been appropriate because he handed me a bowl and filled it with wine. Without thinking of waiting for him, I belted the whole amount in one draught. When I lowered the bowl, the two were staring at me, mouths open.

The old boy quickly recovered and offered me the jug again. I held out the bowl, but this time with a smidgeon of manners I waited for him to fill his. We toasted everyone on the right side of the war. The wine took a good draining for the next half hour. Toward the end the kid even got a bowl.

When the way was clear for me to move on, the boy came along with me for about a week and was a real help. I kept calling him "Hey Stupid," until one day I heard him talking to another person and discovered the word sounded about the same in French. This embarrassed the hell out of me but didn't seem to bother the kid at all. He just grinned, but I stopped using it.

There was no use trying to travel in darkness, so Middleton did his moving during the daytime. His blue RAF pants, a blue peasant jacket, and a beret made him reasonably inconspicuous. Finally, at a village on the main railroad line between Paris and Bordeaux, Middleton entered a tobacco shop, flashed his RAF dogtag—the only identification he had—told the man who he was, and asked for some cigarettes and help (in that order). Middleton's gamble in revealing his identity to strangers at first seemed to pay off, but ultimately failed:

> He hustled me into a back room, sent his wife for someone, gave me a pack of cigarettes, and told me to sit quietly. I figured I'd had it, but I could not think of anything else to do.
> To my surprise and relief, the shop owner arranged for me to stay several days with a family, resting and eating. They even found food stamps for meat for me, so I lived well. I was given a suit of clothes, a passport without a photograph in it, and a bus ticket to a town near the Spanish border—all this without help from the organized Resistance.
> All went well until I reached the demarcation line between occupied France and the nonoccupied region. Here my passport without picture was my undoing. I wound up in jail in Poitiers, in solitarty confinement during Gestapo questioning for three weeks.
> The Germans threatened to shoot me as a saboteur. I told them I was a military man, not an enemy agent. I told them a whole squadron of pilots had been dumped in the same area, and suggested they check their records.
> There were always two of them dealing with me. They came back the next day and said they would believe me if I could give them the first names of the pilots that had been picked up September 26.
> The first name they gave me was Jackson, and I had a sinking spell. I had never heard his first name. I knew him only as Jack, or sometimes as M.E. After flunking the first one I got all the next names right. At last they credited me with being a legitimate U. S. Army lieutenant—and so to prison camp. I think the date was November 9, 1942.

* * * *

Robert E. "Bob" Smith, another Eagle shot down in the flight over Brest, found that luck could be built on logic. Smith said that when the antiaircraft barrage burst around the fighter formation, one plane off to one side and above took a direct hit and blew up. A second round of ground fire sent fragments into his plane, and a small piece ripped through his glove, gashing his throttle hand.

> I broke for the ground, full bore. Another burst exploded below me, jarring the Spitfire. I made it to absolute ground level, dodging buildings

and trees and wires—relatively impersonal threats compared to that sky full of metal chunks and high explosives seemingly dedicated to my personal destruction. I happened to be heading east, and for lack of a better plan just kept on hedgehopping toward open country. Anything was better than providing more target practice to whoever was doing the shooting.

That explosion below me over Brest had apparently been more than a near miss, holing my radiator or coolant lines somewhere. The engine began to run rough and the temperature gauge started rising. Extended flight in any direction was no longer in the books. I was down to about 10 minutes of fuel remaining, and I hadn't the foggiest idea as to where I was.

My keen intelligence cleverly concluded that something was awfully wrong. It went something like this: "If those gunners are British, they need a crash course in aircraft identification. That has to be German flak, so just where am I? Lord, how about pinpointing me geographically and following up with a little divine guidance."

The Lord didn't come through loud and clear, so I was on my own. The alternatives were to belly in gear-up somewhere or bail out. The former made me pucker as I looked at the small fields and thought of all the bad things that could happen—including being unconscious or unable to move near a very conspicuous wrecked Spitfire. The choice was easy to make.

I climbed into the overcast, reasoning that if the Spit dove toward Point A and I fell quietly out of it at Point B, there was a better than even chance that attention would be focused on the airplane instead of on my body.

Smith's logic was on the button. He bailed out successfully; the Germans found the wreckage of his plane but failed to track him down. Thanks to the remarkably clever French Underground, he made his way to a safe house in Toulouse, where he was delighted to find an Eagle downed even before the Morlaix mission, Eric Doorly. "Eric was thin and bright yellow with jaundice," Smith said. "Both of his eyes were bloodshot red because of the negative gravity force at bailout, but he was alive and able to walk."

Eric Doorly had been forced to bail out into enemy-occupied France on September 6, 1942, just 20 days before most of the rest of 133 Eagle Squadron was wiped out farther along the coastline to the west and south.

After taking to his parachute, Doorly had avoided capture by hiding under a blanket of fallen leaves while German troops searched all around him. Late in the night he had crawled to a stable and caught some sleep. His is a narrative of the French Underground Network at work, shuttling the young American successfully to the Spanish border:

> Around daylight a farm hand came into the barn and found me. I had learned some German in school, and was able to understand him and to make myself understood. He was a refugee, hiding out from the Germans, and was working for the family that owned the farm.
>
> I had come down near the village of Aumale, about 35 miles southeast of Dieppe. The farm family was petrified with fear of the Germans. They

gave me a suit of clothes and asked me to move on. I walked all that day, circled around, and wound up in the evening near Aumale again.

At dusk I approached a young Frenchman and got across to him the fact that I was an American. He took me to his home, run by a widowed mother with her two sons. They sheltered me for four weeks while my eyes healed from the stresses imposed by the violent turns of my damaged plane. The negative G's from the outside loop had ruptured the small veins. No pain, but the whites were completely red.

They took my picture and had a false identity card made for me as an apprentice dentist with the first name of Henri. The last name I do not remember. They were wonderful people. The mother picked mushrooms in the fields of a morning, and fixed me mushroom omelettes you couldn't believe. Theirs was a two-family house, all on one floor, and next door a woman and her daughter were entertaining German officers day and night. My benefactors—the mother and her two sons—were away working each day. Once I thought I was seen by two German soldiers coming up the walk to visit the two women next door. It was a real scare. Those were interesting days.

I was alone all day and used to listen to a German radio station's English-language broadcast to pass the time. On September 26, 1942, the Germans announced the names of 11 American RAF pilots who had been shot down or crashed on the Brest peninsula. It sounded like the whole 133 Squadron roster and almost was.

After a month the two boys told their mother they wanted to go to England and join the French resistance forces. She agreed, although this meant she would be left all alone. We took the train to Bordeaux in southern France, walked over the line of demarcation, and then went by bus 130 miles south to Lourdes, where the boys had an uncle.

By the time we got to Lourdes, at the northern base of the Pyrenees, I had turned yellow with hepatitis. A German refugee doctor treated me, and I consumed a lot of water from the famous shrine. My friends there would give me enemas and then stand around to observe the results. They were happy when I was successful.

After I had been there about three weeks, the French Underground contacted me and took me by bus to Toulouse, in the part of France not yet occupied by Germans. Here, in a huge house with six or eight bedrooms, I met up with Bob Smith. He had been on the most disastrous Eagle Squadron mission ever flown, in which 12 new Spitfires were lost, 20 days after I had to hit the silk. Like me, he had managed, with the help of friends, to evade the Germans.

Doorly generally appreciated his chance to recover in Toulouse:

The house was a major headquarters for the underground. I have never before or since seen so much money. There were 1,000-franc notes printed in England, with pin-holes put in to make them look authentic. One man even had a short-wave transmitter, here in the middle of the city of Toulouse, and was operating it to transmit espionage data to England. We ate like kings on the black market. Bob and I slept in the same bed, and that was a mistake. I caught scabies, an infectious skin disease, from him. He caught it in a much more enjoyable way.

But there was yet more ground to cover. Eric Doorly continued:

A plan was underway to get us out by submarine. While the arrangements were being made, the Germans came in and occupied the rest of France—and that killed this option. A new scheme was devised; we would walk over the Pyrenees into Spain. More time now must be allowed, to set up guides for the trip.

We boarded a train and set out for Pau, a town of 40,000 close to the Pyrenees, about 20 miles north of Lourdes. There were Smith, myself, a couple of British fliers—five or six escapees and five or six guides in all. It was quite a train ride, up close along the border. We were in a typical compartment, six people seated on each side.

A Vichy French official came in and started to check our papers. Immediately, there was a tremendous conversation going on among all of us, except for me. I could not speak a word of French. I was sitting in a corner and was the last man to be checked. The official looked at my ID card but did not say a word. There must have been something that I did not know about.

Ten miles from the border the train slowed almost to a stop. We got off and started walking. It was December and very cold. It took us two days to get across the mountains to the frontier.

The first time we crossed into Spain we were picked up and taken back to the border and left there. We shifted a few miles to the east, crossed, and got caught again. Finally a guard agreed to take us to Barcelona if we would pay his and our bus fares.

Smith and another man had split off earlier and walked all the way into Barcelona. I rode. Scabies made my legs look like raw meat, and left me too weak to hike. In Barcelona our party, minus Smith and his friend, ended up in a prison. We were in a cell 8 feet wide and 15 feet long, 30 or 40 people with just a hole in one corner. There was room to sleep if we lay on our sides, but not if we lay on our backs. We could get a cup of coffee and a piece of bread in the morning, another cup at noon, and another at night. It was a pretty rough place, filled with murderers and other desperate criminals, along with a lot of pretty bewildered refugees. But if you had money you could buy anything, from women on down—or up, whichever way you thought of it.

We were there a month until we were tried for clandestine crossing of the border and for bringing in money and were sentenced to a concentration camp in Northern Spain. We were taken in boxcars to the camp at Miranda de Ebro, a godforsaken place, but not too bad compared to the Barcelona prison. There were several horribly unsuccessful escape attempts. Two fellows tried to get out through sewer pipes. They got stuck in it and drowned in the sewage mains. Many of the inmates had been on the communist side in the Spanish Civil War. Some had become camp guards.

* * * *

While Doorly's journey was taking him to one undesirable destination in Spain after the next, Bob Smith's fortunes were comparatively good, despite similar hardships. Near the end of the rail trip with Doorly and others, Smith decided to try it the rest of the way on his own. He jumped out of the slow-moving train and started to climb the Pyrenees. In extreme cold, he reached a cabin containing 10 other fugitives, one of them a woman.

Smith and three of his new companions—a Wellington bomber pilot named Barnard, a sergeant, and a Pole—elected to struggle on. The rest remained in the cabin. The hardy quartet made it across the border into Spain but were picked up by the brutally hostile *Guardia Civil.* They were held first in a square hut in Barcelona without heat, water, or plumbing and then were transferred to a tiny room in the city jail. After a week, Bob Smith got word to the British Embassy and was released to a resort at Zaragoza to await transportation back to England.

Eric Doorly, after two months at Miranda de Ebro, received word that he, too, was getting out. As he recalled, "When the happy day arrived, eight of us were taken out on a bus with a Spanish lieutenant and a driver. The Spanish officer was a real nice guy, and it was a great trip. We got some booze and drank it. Then we talked him into stopping at a whorehouse. We paid his way, and everybody had a good time. We wound up at a hot springs resort area. There were no other guests but us, and we stayed there while final arrangements were made. Then we proceeded by bus to Gibraltar. Spain was a neutral country, and technically we should have been interned for the rest of the war. British influence was strong enough, however, to get us out.

"When I was shot down, I was wearing soft boots, lined in fleece. After all my walking, the soles were flapping. In Gibraltar I could find no size 12 shoes, so I arrived back in England, after an absence of seven months, wearing new sneakers."

Doorly wryly characterized his September 1942 bail-out into France as a bit of bad timing because it came only days before the transfer of the Eagle Squadrons into the U.S. Army Air Forces. "My pay as an RAF flying officer was $80 a month," he said. "As a U.S. Army captain it would have been $300 a month."

9
Eagles Caged in Sagan

DOORLY AND SMITH counted themselves quite fortunate to have evaded the Nazis and to have fled from the Spanish jails. For 26 fallen Eagles, their lot was the ever-swelling Stalag Luft III—Air Camp 3—in lower Silesia. This camp eventually held over 10,000 allied officers within its four compounds, and was to remain the main prison camp until the winter of 1944–45.

The original compound there held British POWs. The Americans captured before the first USAAF raid of July 4, 1942, all were in RAF uniform, and therefore were locked up with the British.

During the winter of 1942 and into 1943, the Americans became fully operational, and their losses began to mount. The Germans, using information and experience gained with the original camp, built a supposedly escape-proof North Compound in the vicinity and moved the Eagles into it. As of March and April 1943, American pilots, still billeted with the British, for the first time had a line of barracks of their own. In the summer of 1943, the Eagles were further transferred to the South Compound, where virtually all the inmates were American airmen, and in March 1944 they settled into the all-American West Compound.

Their lives in Sagan were described later by some as "sheer boredom devoid of really severe treatment, punctuated by periods of great hunger." But the experiences of those caged Eagles also reveal their resourcefulness and camaraderie and, despite the odds, their relentless and instinctive desire to break loose and fly free once again.

For Bill Geiger, going to Stalag Luft III was "the beginning of three and a half years where every day was like a week, every week like a month, every month a year. Being a prisoner of war can be described as sheer boredom. You never were really sure when it was going to be over, and if you were going to be around when it was over. We went through periods when we were very hungry. We were cold in the winter; hot in the summer.

> One of the things that makes it possible for a person to survive as a prisoner of war and to maintain sanity is the liberal use of a sense of humor. This is assisted because you lose your reference point. You are all there living under the same conditions, suffering the same discomfort, and there is no reference point with home any longer. Even the mail when you get it seems to be a rather distant thing.
> It is a little like flying in an airplane. There is no reference point that ties you at 30,000 feet to the ground. As a result you do not feel the height. You have no sensation of height or speed. This same thing happens to you in a prison camp.

Geiger and the other Eagles became well known to then-Lieutenant Colonel, and later Lieutenant General, Albert P. Clark, a 1936 graduate of the U.S. Military Academy at West Point, New York, who described himself as "one of the first three or four U.S. military types captured by the Jerries." Clark was second in command of the 31st Fighter Group, the first U.S. Army Air Forces fighter group to reach the European war theater. The Group arrived in England in June 1942 and was equipped with British Spitfires.

"I was shot down in Pas de Calais on 26 July 1942 while seven of us were learning the trade flying sweeps over France with the RAF Tangmere Wing," Clark wrote in 1984 in response to a query. "I was flying with a Canadian fighter squadron (412) and was shot down on a sweep over Abbeville."

Clark arrived at the Sagan camp from Dulag Luft about August 15, 1942. Ironically, he took the bunk recently vacated by none other than Bill Hall's ex-roommate at St. Omer, Douglas Bader, who, Clark said, "was removed by the Jerries as a trouble-maker and sent off to a straff lager at Colditz."

Clark remembered the skinny Bill Geiger:

> By September we had a dozen or more Americans, including the Eagle types, who wanted to room together. The first American Room—eight of us—included Bill Geiger. We called the Room "Little America" and lived together until we moved to the new North camp in March of 1943.
> I got to know Bill quite well. He had a delightful sense of humor and made clear his plans after the war to marry a rich and lovely lady and live in idle luxury the rest of his life.
> Bill was as thin as a rail, and we were worried that he might be tubercular. One day we acquired a lemon—how I don't remember as it was the only one I saw in three years. The Room voted to give it to Bill for the vitamin C. He ate it skin and all!

Lieutenant Colonel Clark was also at the gate in early October 1942 to meet George Sperry and Charles Cook, downed in the Morlaix mission fiasco, and he was later to get to know and even admonish the irrepressible Bill Hall.

Although the German Luftwaffe controlled the overall camp, each compound at Stalag Luft III was run like a military base by its own officers. Sperry and the other Eagles started off their term in the East Compound, quickly learning a few practical tips from the experienced British POWs they encountered there.

> The old hands showed us how to vary the daily menus a bit even though the basic ingredients—largely from the Red Cross food parcels that brightened our lives—almost never changed. The experienced camp inhabitants taught us the art of making cooking pots and pans from salvaged tin cans. They helped us adjust and endure. From them we learned how to cope.

To relieve boredom, Sperry said, most of the caged Eagles volunteered for spare duties. "I worked as a security officer in training under the British in North Camp, and became security officer working for Lieutenant Colonel Clark in South Compound. I was a block commander, and not only hid tools in traps but covered for the factories working on escape activities."

* * * *

In the prison camps, the POWs were known generally as *kriegies*, short for the German word for prisoner of war, *kriegsgefangener*. But perhaps because of his seniority among the Eagle prisoners or his popularity as an unusually genial and friendly individual, one tall young man acquired kriegie as a nickname: "Kriegie" Bill Hall. He recalled the Eagles' life in the East Compound.

> At first Stalag Luft III had only one compound, which eventually became known as East Camp. It was made up mainly of RAF POWs. Wing Commander Bob Tuck, the British ace who shot down 32 German aircraft, was a prisoner there. The Canadians and Americans were just starting to come in. It was a fairly quiet camp, but anyone caught out in the yard at night would be shot. The cement garbage burners provided some cover until a man could sneak back into barracks.
> Escaping was a serious business in East Camp. We had a head escape master, a man in charge of civilian clothing, a man in charge of escape rations, a man who was good at passports and German I.D. cards. Others were good carpenters and could move walls out and so forth. The East Camp was a good spot for tunnels, as the abort was quite close to the fence with woods on the other side. If you went to the toilet and happened to sit close to the wall, somebody might touch you on the bottom and ask you to move, as they were working down below.
> At the end of March 1943 I was moved to a new compound, which became known as the North Camp. East Camp by this time was becoming very crowded, so late arrivals were being sent to North Camp. These included a few Englishmen and many Canadians and Americans who were coming in fast from the shot-down B-17s.

The North Compound was designed to be escape-proof. Given the history of tunnelling activity in the first compound, the Germans picked the North area to house POWs because the sandy soil there made tunnelling almost impossible. But the very process of clearing the North area of brush and trees was to give the Eagles their first, albeit bittersweet, taste of freedom.

Barry Mahon recalled the origins of the "tree trunk double escape:"

> As we entered this so-called no-escape camp, we saw trucks still hauling out the young pine trees that had been cut down to make our athletic field. Leroy Skinner and I immediately requested permission from the X Committee—the prisoners' secret escape planning body—to try to get out by climbing on top of one of the buildings and jumping down onto the flatbed of a truck as it went by.
>
> The X Committee had a rule that when someone discovered an escape route several other people should try it out first. The first two men to try it were successful and went out the gate, unnoticed among the tree branches in the truck. George Middleton and M. E. "Jack" Jackson were next on the list, ahead of Skinner and me.

Jackson described what followed:

> The new area had been heavily forested, and part of it was being cleared for a parade ground. Trees and brush were being cut down and hauled out on trucks.
>
> I had noticed that Russian slave laborers were building a road so that the trucks had to detour and pass right along the side of our barracks. So Middleton and I figured we would escape together by climbing on top of the barracks and jumping down when a truck came by.

To simplify matters, the two men agreed that Middleton would carry maps and a compass concealed in a cigarette packet, and Jackson would carry money, provided by the X Committee. Middleton knew some French and German, and would do most of the talking whenever communication was necessary.

The men clambered to the barracks roof, as planned. Jackson made a graceful flying leap and sank out of sight in the branches. Middleton jumped, but was too far back. He missed the truck and sprawled on the ground. It might have been a comical scene, but none of the prisoners watching felt like laughing. Middleton's plight was too desperate.

> Jack jumped off the roof first. In order not to come down on top of him, I landed just a hair too far to the rear, whistled through the branches and onto the ground.
>
> I strolled into the barracks as calmly as I could, but that supposedly nonchalant stroll was about as hairy a thing as I have ever done. I expected to be shot down at every step. You'd think two dummies jumped off the barracks all the time, for all the lack of attention I got.
>
> I went back up on the roof and did it again, but of course now Jack

and I were separated. I had the maps and Jack had the escape money. I could speak a little French and some German—so little that the only good it did me was to make the road signs easier to read.

"We never did get together," said Jackson, taking up the narrative.

> I wandered around without map or compass for several days, not knowing what way to go. Finally I got to Frankfurt an Oder and went into what I thought was a church, hoping to find a sympathetic priest. It turned out instead to be a manufacturing plant.
> That was the end of my freedom. I had been out 8 or 10 days. My punishment was the standard 14 days in solitary, on bread and water.
> Middleton crossed the Oder River into Poland, but was caught there.

Back behind walls and barbed wire after his brief taste of freedom, Jackson headed up the American escape unit in company with Lt. Col. R. M. Stillman, the Compound Executive Officer, and a senior officer, Col. William L. Kennedy. His sophisticated efforts to help POWs escape later won him the U.S. Army's Bronze Star.

> I was considered the camp's escape expert.
> I developed several new ways for packaging escape material, such as compasses, money, and maps. We would hide it in cribbage boards, in cigarette packages. We'd sew cloth maps into clothing.
> Every part of the escape packaging in our compound went through screening successfully. We didn't lose a thing to the German inspection people.

In the meantime, as Bill Hall described, the North Compound became an occasionally boisterous spot.

> North Camp filled up very rapidly and soon we were up to 2,500 people, six to a dozen in a room, from all walks of life—miners, accountants, engravers, anything you wanted. We had classes going in mathematics, esperanto, self defense, and so forth. Bill Geiger provided fencing lessons.
> As soon as we could, we got wooden kegs from the Germans and with the prunes, raisins, and sugar from our Red Cross parcels, almost all the rooms had a brew down. A brew took about 14 days. In 1943, with so many Americans in the North Camp, we decided to have a winefest on July 4. After the 9 A.M. *appell*, or roll call, that day the whole camp started drinking, and by 2 P.M. 90 percent of the camp was drunk and wandering from room to room with mugs of wine. Then some of the boys tried climbing over the fence to freedom. This day the guards were good. They would let the boys climb over the fence and go and get them and lead them back in to the compound. That night there was no appell. After that hectic day with the drunken POWs, the Germans put a ban on wine-making, and from

then on we had to pay off the German ferrets with cigarettes and chocolate in order to keep a brew down in our room.

The constant preoccupation, however, continued to be a hope for freedom, Hall said.

> At North Camp we were all working on some way to escape, and tunnelling seemed to be the best. But the tunnels had to be shored up all the way with bed boards to keep the sand from caving in. One of the problems was to get rid of the dirt from the tunnels. We solved this by wearing an extra pant leg inside our trousers. We would fill this with dirt from the tunnel and go for a walk around the perimeter and, with a trip string, drop the earth. There were always lots of tunnels. The Germans found some with their 10-foot steel probes, and they missed some.

Barry Mahon and Leroy Skinner, who had been designated by the X Committee to follow Jackson and Middleton out on the next tree trunk escape, found their plans frustrated, since the Germans promptly stopped hauling trees when the first Eagles escaped. Mahon stated that this disappointment only postponed his flight for freedom.

> Skinner and I were given priority to be included in the very next escape of one kind or another. At this point Wally Flood, one of my roommates in the first camp, had been appointed head engineer on the tunnels made famous in the book and movie about *The Great Escape*. He asked me to work with him. Since I had no previous experience on tunnels, I was put into the dispersement section. This meant carrying sand that was dug up during the day—carrying it in pockets designed out of sleeves of overcoat lining fastened inside your pants legs. As you walked around the camp you pulled a little string, allowing the new sand to be mixed lightly with the old.
> These three tunnels, if I remember correctly, extended a total of 300 feet. They were 25 feet underground and were at least 2 feet square, plus several collection and dispersal and air pump chambers. I never went into the tunnels myself.
> While plans for the tunnels were progressing, the X Committee developed a special project, mainly to ridicule the Germans and their ambitions for an escape-proof prison. Thirty of us would walk right out the front gate.
> On account of our previous disappointment with the trucks and trees, Skinner and I were approached and included in a so-called delousing party. The plan consisted first of discovering a louse in one of the American barracks, and then putting into motion the German rule that the whole barracks must be deloused immediately. This meant that we prisoners had to be marched under guard out of the compound to delousing showers some distance down the road. I am not quite sure where they found this louse, but it was planted in the bedclothing of one of the boys and promptly shown to the Germans.
> Before this, through much careful planning, two German uniforms had been manufactured using RAF material covered with talcum powder to

make it a proper gray color. Certain pieces of ribbon, piping, and insignia had been collected from German enlisted men under the guise of souvenir swapping. From these originals, casts were made out of cigarette-pack tinfoil that had been collected by the thousands. Enough lead or aluminum was melted to pour into the molds and make genuine-looking buttons and insignia.

A Polish woodworker also had made two replicas of German rifles, down to the detail of an actual working bolt and breech mechanism. These wooden models had been colored by shoe polish and absolutely could not be distinguished from the original.

Preparations went on more or less behind the scenes. Because the escape was to be accomplished in our own uniforms, we were not required to go through much briefing of our crew.

On the day the louse was discovered, all preparations had been completed. Consequently it was arranged that the delousing parades would start the next morning with the whole maneuver tied to the hour at which the sentries on duty at the main gate changed shifts.

Escorted by two German guards, our first delousing group left the camp on schedule with belongings tied up in sheets and blankets to be fumigated. As planned, this group of prisoners would be docile and obedient, with no attempt to escape, in order to lull the German escorts into a false sense of security. The trick now was to delay the new set of guards, coming into the camp to escort the second delousing group to the showers, long enough to involve a new set of gate sentries just coming on duty. We members of this second POW group were the ones assigned to make the actual escape.

As the German escorts for our second delousing group arrived in camp, our adjutant took them into a room across the compound, saying that the delousing group was not quite ready and would the escorts like some good American coffee, compliments of the Red Cross, plus a piece of pie and maybe a cigar. These items were irresistible to the Germans. They had not seen real coffee or cigars for four or five years.

When the sentries on the gate had changed shifts, the two escort guards were informed that their little party was ready and would assemble shortly. In the meantime two of our German-speaking prisoners—one was Norwegian, I think, and the other a Czech—had dressed in the fake uniforms. They led our delousing group right out through the main gate. A pass was involved somehow. I can't remember the exact details, but I think one of our boys managed to tear a blank pass, or several of them, off a book of passes and stuff them in his pocket. At any rate, a pass was made out for this number of people.

Since the sentries now at the gate, having just arrived, were not aware of the German escorts who were enjoying coffee and pie with our adjutant, we walked past them and out of the camp without a hitch. As soon as we were clear, and well down the road, our fellow prisoners still inside the camp assembled the real shower parade and called the two real German escort guards out of the coffee klatsch. Now again the gate sentries were a bit confused. Their records showed that only two escorting guards had come in the gates, and now the number going out would be four. But since the guards coming out at this point were legitimate, the new sentries marked the mix-up to an oversight on the part of the previous sentry team, and they allowed this parade to leave in the direction of the showers.

Meanwhile, our group had continued down the road just far enough to get out of sight. Instead of going on into the German part of the camp for our showers, we broke and headed for the woods, shepherded by our two phony escorts. Not until evening, when the books at the gate still showed one shower parade missing, did it dawn on the Germans that 30 officers had walked out through the main gate of what they had considered an escape-proof camp. There were 10,000 Allied officers in the camp enjoying the biggest laugh they had had since they had been in Germany.

Skinner and I immediately changed our uniforms to those of French workers. Not much of a change; merely putting on a beret and wearing a French overcoat. We continued walking south that day and on through the night through a reforestation area, and actually saw no one for the first 60 or 70 miles. We headed toward Czechoslovakia and eventually followed a railroad that our map indicated would take us in the right direction.

The next day about noon we came upon a quarry being mined by British enlisted-men war prisoners. We practically walked into the middle of this thing before we realized where we were. We turned and went slowly up the hill, hoping that no one would notice. Once on top of the hill, there was no exit. We were so tired we dropped in our tracks and slept several hours. We were awakened by a gentle rain and looked down to find the quarry empty.

This much concentrated walking had caused both of us to develop all kinds of foot ailments, including very deep blisters. After this deep sleep we found that our muscles would not operate at all. Helping each other to a sitting position, and eventually standing, took 15 to 20 minutes of sheer pain. About the only way we could get started was to propel our bodies forward and let self-preservation make the muscles work. After a while the kinks were out and we were walking normally again.

We were cutting through a wheatfield when we heard voices almost directly in front of us. We dropped in our tracks, and soon figured out it was a German man and woman dallying in the green. He was giving her every phony argument that any soldier ever has used to overcome her resistance. The seduction took several hours. It was hard for us to contain our laughter. When finally the two had left, we continued on across the field and picked up a road that we believed would take us into Czechoslovakia.

This particular evening the moon was out and the walking was easy. However, because of our fatigue, we stayed to the road instead of cutting through fields as we should have done. As we got to a settlement near the border, we found ourselves unexpectedly inside the village, with no chance of turning back. There were several guards along the road, in the German custom. We had passed two of them and were almost clear of the town when a third guard, quite elderly, stopped us. We told him we were Hungarian workers, but he took us to the Burgermeister, the mayor of the town. There they searched us and found our escape rations, our dog tags, our map.

The whole police network of Germany had been alerted about our escape, and the old guard who captured us was an instant hero. I imagine it would have been easy for us to have pushed him into the river. He was armed with a giant revolver that probably would have exploded had he tried to fire it. But our orders were to get out as cleanly as possible, without violence, in order to maintain our status as escaped POWs, rather than wanted criminals. Escaped prisoners were returned to their establishment, while criminals would be taken by the Gestapo or the police.

We were driven to the next village, a large railhead. Eventually two guards came to get us. Since it was an evening train trip, they were allowed to bring their wives. On the return to Sagan, the conductor locked the four of them and two of us into a compartment.

One of the women spoke a certain amount of English, and Skinner and I knew just enough German to carry on a halting conversation. The talk evolved around our camp, which the woman said must be Utopia, full of sugar, coffee, cigarettes, and other luxuries. She said she could not understand why we should want to escape when our lot in camp was so much better than that of the average German civilian.

Not wishing to go into the political or military considerations, I told her I wanted to escape because I missed my mother very much. That was true, of course, but not on the exact sentimental scale that I had sounded. Germans of this middle class are very family-conscious. As my remarks sank into her mind, tears came into her eyes and she began to deride her poor husband. "It is shameful that these fine young men are being held in camps," she said. The husband looked at me and shrugged his shoulders and said in German, "Why did you have to get her started on something like that?"

We were taken back to camp and given 14 days of solitary confinement.

* * * *

Eagles were involved in the planning and execution of other escape attempts at Stalag Luft III. In the 1970 book, *Free As A Running Fox*, Squadron Leader Thomas Calnan, an Englishman with a penchant for escape techniques, recounted one plan, involving Eagle Morris Fessler, that had to be abandoned.

Calnan hit upon the idea of trying to smuggle himself out of prison with an RAF group that the German guards would twice a week escort out of the gate to a gymnasium on the far side of a moat. Since the guards would not know he was in the party, Calnan thought he might be able to hide in the moat until darkness would enable him to get away.

> I spent hours establishing that it was impossible for me, a small man, to hide myself under Fessler's Polish great coat. I tried hanging upside down from his shoulders with my feet tied together with a rope that bore down on his neck. Even though Fes was a very big man and his Polish greatcoat was an immense shapeless coverall, it did not work. As he lurched along with me suspended under this coat, he looked so enormously fat that the guards would certainly have been suspicious. We had to throw that idea away.

Fessler, however, played an important role in the noteworthy escape attempt by Bill Nichols and a companion that was characterized by Bill Geiger as "probably the most courageous I ever saw during the war." Fessler described the planning of this flight in broad daylight:

> I noticed, while walking the perimeter of the Sagan compound, that the coiled wire inside the double fence was a big obstacle to vision, both to the guards in the look-out towers and to the guard who paraded back and forth between each tower. It looked to me that these guards, even though directly above it, perhaps 25 to 30 feet high, couldn't see into the coiled barb wire more than 100 feet or so laterally. The same would apply even more so to the parading guard. The towers were approximately 300 feet apart. That gave an approximately 100-foot area between the towers that was blind to the tower guards and would be blind to the parading guard as long as he was about 100 feet away.
>
> The plan required timing each parading guard very carefully and noting each guard's habits, selecting a day when a particularly slow parading guard was on duty. It also called for providing other visual diversions for the other guards across the other side of the compound, and indeed for this particular guard if found necessary.
>
> The object was to have the tower guards and the parading guard's attention momentarily diverted while two escapees went the 20-foot distance from the guard wire to the double stranded (coiled barb wire) fence—a matter of only 4 to 5 seconds. Then the escapees would be lying next to the coiled barb wire and out of sight of the tower guards and the parading guard, providing he was in the proper sequence and position on his parading schedule between the look-out towers. It was then necessary to cut through the double stranded fence and the barbed wire rather quickly—a matter of 3 to 4 minutes at most. It was necessary to get through the fence and out the other side before the parading guard turned around and again faced that direction, or he would see you come out of the fence. Once through the fence, the two escapees would stand up and leisurely walk away. As this plan was being finalized, I was shipped out of this camp to Shubin XXIB in Poland. Another chap, an Irishman—Ken Toft—took my place in the scheme, and the escape actually took place and came off exactly as planned in broad daylight!

Bill Geiger was a witness to the Nichols/Toft adventure:

> They actually, in broad daylight right under the guards' noses—with a little help from other POWs from a diversionary point of view—jumped right over what we called the dead-man wire, crawled right up against the fence, cut their way through the fence, crawled through it, and got out on the other side—and calmly walked off into the woods like they were a couple of French laborers.
>
> It was a very gutsy thing to do. They weren't out very long—two or three days, at the most. It was not very much of a reward for that kind of courage.

* * * *

Outside the camp, the Eagles' air war turned increasingly to bomber escort service. Oscar Coen, the original evader, continued to play a major role in the action. On October 2, 1942, now USAAF Capt. Coen took part in the first combat mission of the 334th Squadron, a fighter sweep into France, and shot down an FW 190. Two other squadron members, Gene Fetrow and Stan Anderson, also shot down enemy aircraft. James A. Clark, Jr., and R.M.B. Duke-Woolley shared in yet another kill. On January 22, 1943, Coen, Stan Anderson, and Robert Boock each downed FW 190s while escorting Boston bombers to St. Omer.

Early in 1943 the former Eagles, now the key pilots of the U.S. 4th Group, started turning in their aging Spitfires for new and largely untried P-47D Thunderbolts. Coen, newly promoted to major, was among the 11 squadron members taking part in the first P-47 combat mission on March 10, a sortie into Holland. "A lot of German fighters came up, but they just stood off and looked at us," pilot Don Nee reported. "They hadn't seen Thunderbolts before and were wary."

Coen's new Thunderbolt betrayed him April 3, 1943. "A couple of cylinders cut out at 29,000 feet and there was an explosion," he said. "This was before we had barometrics to open our chutes and before we had walk-around oxygen bottles. I had to bail out and had to open the chute right away, although in that rarefied atmosphere the airplane's speed was around 350 miles an hour."

> I used my right arm to pull the ripcord, and the sudden force fractured and dislocated it. I almost lost my arm. Luckily, I was over England. I gave myself a morphine shot in the thigh on the way down and passed out. I landed close to a field hospital. A rescue team was there within five minutes. I recovered consciousness as they put me on an ambulance, and told them about the shot of morphine.
>
> I had heard of a couple of other pilots who lost the use of an arm in high-altitude high-speed bailouts, but I was fortunate enough to have friends, especially Doc Arthur Osborne, our medic and chaplain from Georgia, who put me on exercise therapy right away. Thanks in good part to that treatment, I missed only two months of duty.

On May 21, 1943, Hawaiian Eagle Brewster Morgan undertook a significant mission that he mistakenly hoped would be "a milk run."

USAAF B-17 bombers were to make their first penetration into Germany to Emden, just across the Dutch border, and P-47s of the U.S. 334th Fighter Squadron, successor to the RAF's 71 Eagle Squadron, were to try to divert toward Brussels any enemy fighters that might threaten the bomber formation. "Our P-47s were too short of range to escort the bombers all the way, so we had to resort to the tactics of diversion," Morgan explained in a December 1977 article in the *Honolulu Advertiser*.

After the P-47s had crossed the English Channel, Radar Control in England warned the pilots that "30 plus bandits" were taking off from Antwerp. In the furious combat that followed, Gordon Whitlow, flying a plane with code letter QP-H, and Lee MacFarlane in QP-N, were shot down and killed, and Morgan's plane, QP-U, was set on fire. The squadron report on this mission observed, "Coincidentally, letters of missing planes spell out HUN." Morgan said he sent one of the attacking Messerschmitt 109s into a steep dive with machine gun fire, and his wing man, Bud Care, sent another Me 109 falling away out of control and on fire.

> The hits that set my plane on fire were in the engine and apparently behind the cockpit. When I tried to pull the canopy release, it would not slide back. In the C-2 model P-47 there was no way to jettison the canopy; you had to depress a release on the forward part, and the canopy could then be slid back. It was my opinion that my canopy was riveted shut.
>
> I dumped my shoulder harness and seat belt combination and again attempted to open the canopy. My plane was in a spin most of the time, and each time I pulled out of the spin the Me's would hit me again.
>
> The 109s attacked once more. This time a shell exploded in the cockpit. A splinter struck me under the left eye, inflicting extreme pain. As blood spilled down the front of my left vest, a bullet smashed into my left leg just below the knee. The tail assembly had been damaged and the ammunition doors blown open. The only instruments still readable were the airspeed indicator and the altimeter. At 400 feet above the water, wallowing along at about 200 miles an hour and nearly out of control, I finally pried the canopy open using a small steel crowbar I carried in my boot at the suggestion of RAF pilots who had been trapped upside down in crash landings.
>
> By now I was too low to bail out. Unable to strap myself in again—my shoulder harness had slipped behind my seat—I decided to stand up in my seat to land the plane. I pulled back on the stick as hard as I could, and the plane, rather than nosediving, fell almost parallel with the water. It hit with a tremendous jolt that threw me high and clear. I landed on my back, with the blow buffered somewhat by my parachute and the life raft curled over it. I sank immediately, but pulled the gas relief valves on the life vest and shot to the surface. Four Me-109s circled above and waggled their wings as I pulled myself into my life raft.

Still afloat as night fell, Morgan was drifting steadily toward the enemy-occupied Belgian coast. Fingering his wounds, he pulled a quarter-inch-long piece of shrapnel from under his left eye. His other injuries appeared to be superficial. Morgan reflected on these facts: the Germans must have mined the beaches of this area; the 50-mile security zone extending back from the Channel coast virtually ruled out hope of a successful walk-out; the bullet in his lower left leg made any kind of a hike unfeasible.

> I decided to fire one of the flares in my escape kit. Maybe friendly Dutch or Belgian civilians would find me and help me. A moment after sending up the flare, a bright floodlight turned on me, and I heard people speaking in French. Within moments I was being pulled aboard what appeared to be a Belgian fishing smack. My rescuers placed me gently on a pile of rope. In my schoolboy French I asked if they could help me escape from the Germans. Their answer: *"Les Alemans viennent."* (The Germans come.)

I discarded my escape money and my pistol. Shortly a German E-boat arrived. I transferred to it with a young German naval enlisted man putting a pistol to my face. "For you the war is over," he said.

Ashore I was driven in a commandeered truck to a naval dressing station, where a young navy doctor examined my wounds. He was very much anti-Nazi and spoke perfect English, having been stationed in England. He dressed my face wounds tenderly. An aging enlisted man fed me the first orange I had seen in over a year, and he cried at my appearance.

About an hour later two Luftwaffe sergeants escorted me to a staff car and we drove to a small field hospital, where a medic put stitches in my face. The doctor apologized for his inability to give me an anesthetic, explaining that there was a shortage of medical supplies because of the heavy demands at the battle fronts.

Next day they took me to a Brussels military hospital. I remained there for about two weeks, until I tried an escape from the men's room. Next they sent me to Dulag Luft for the usual interrogation—no military questions, just queries about how we lived and about our war experiences. They mentioned that a P-47 with an H on the side of the fuselage had been found near Bruges, and that is how I learned that Whitlow had been shot down. There was a surprise visit from Major Joseph Priller, an ace with 101 victories, and his flight commander, a Hauptmann Schuman. "We think we shot you down," Schuman said.

Morgan became an inmate of Stalag Luft III and soon found that the British were running an effective communications network. He reported to the senior British officer of the camp, Wing Commander Harry M.A. Day, who had been in German prisons since 1939.

Day questioned me closely as to any air bases I might have observed on my trip to Stalag Luft III. I had noticed a large Folke-Wulf aircraft assembly plant near Dresden. The next week a large force of B-17 bombers raided the plant. I found out later that the British had a means of notifying their headquarters in England of this type of intelligence.

* * * *

As the bombing escort missions grew in number, American fighter planes also needed greater range to fly support deeper into Germany. In July 1943, pilots of the USAAF 335th began flying P-47s with additional external fuel tanks in order to carry out more extensive missions.

Hank Ayres was shot down on July 28, 1943, while on an escort mission for B-17s into Emmerich, Germany.

Operating as Section Leader, I was forced to turn back because of sudden engine trouble. No one came with me, and at 1,600 feet five Me 109s attacked. I tried to evade them by out-diving, but when I reached the deck they were able to quickly overhaul me because my engine was working very poorly. I was finally shot down over the town of Gouda, Holland, and got a compound fracture of the left thigh.

Thereafter, he was held in hospitals at Gouda, Amsterdam, Frankfurt, and four other German-held sites. He would ultimately reach Sagan nearly one year later—and then be repatriated through Sweden in September 1944 in an exchange of wounded POWs.

On August 16, in a record-setting day for American forces, the 4th Group squadrons shot down 17 enemy aircraft, damaged five more, and reported one probable kill during an attack on the airdrome and repair installations at Le Bourget, near Paris. One pilot, Joseph Matthews, formerly of 121 Eagle Squadron, was shot down, but managed to evade capture and rejoin the action within a few weeks.

In September, 1943, two more Eagles came into Stalag Luft III. On a September 7 mission to Brussels, Lieutenant Aubrey Stanhope was shot down. Two days later, Lieutenant Frank Fink, also a 121 Squadron alumnus, was shot down during a raid on Paris.

* * * *

By September, the already caged Eagles has been transferred from North Compound to the new South Compound. Charles Cook changed over in June 1943; Bill Hall transferred at the end of August. The move split some of the Eagles apart. Danny Daniel, one of the transferees, stated that he left behind fellow Eagles Fessler, Maranz, and Nichols, and only saw them through the fence for the remainder of their confinement.

Hall described the South Compound experience generally, and his specific encounter with Lieutenant Colonel Clark, as follows:

> At South Camp the routine was the same as it had been at North: 9 A.M. roll call, and another at 5 P.M. or just before dark. Sometimes the appell lasted five or six hours, and the Germans would keep us out on the parade ground until the barracks were all searched.
>
> At the beginning, Bill Geiger and I shared a small room with just two bunks in it and a stove and table. We were given only about three bricks of coal a day for the stove, for heating and for cooking, and were always cold. Later, as the camp filled up, we shared our room with two others. The routine was pretty much the same at all the camps. Aside from the two appells a day you just did things to keep yourself going and as sane as possible.
>
> We sat around and did a lot of talking, played cards, horseshoes, anything that came into our heads. We got lots of exercise on the horizontal bars we had made. Bill Geiger was a pretty fair fencer. Once he and another lad staged a diversion for an escape by having a fencing match out by the main gate. The fellow who made the escape had fashioned himself a German officer's uniform, and he simply walked out the gate while the fencing match was going on.
>
> As well as being adept at fencing, Geiger was good at throwing gramophone records after we had heard them a million times, using a fast spin that would leave pieces sticking in the wall. The wax records also came in handy in other ways. We made compass cases out of them by heating them in hot water and then putting them in a form to make a compass case about 1¼ to 1½ inches across. Then we added the magnetic nee-

dle and put a glass window in the top of it. I produced a few good waterproof compasses using pitch from the pine trees for a sealer.

We also used the pine pitch as a cleaning solvent for solder. We got the solder off old tin cans and ended up making ourselves stills.

Bill Geiger, John Dorch Lewis, and I were able to double-distill our wine and make an alcohol so strong it would burn a blue flame in a spoon, and burn dry. We did our distilling at night in the kitchen when everyone was asleep, and kept our still hidden in the attic when we weren't using it.

On this one occasion we had been distilling all night long and had just finished running the alcohol through a second time when the German ferret came in. It was early in the morning and we were just hiding our still up in the attic. He saw what we'd been doing so we told him to have a drink. We poured him a drink of the alcohol mixed with orange concentrate and water. The ferret drank the first drink and we told him to have another, thus giving ourselves time to hide our still.

We were planning to have a drink ourselves, but when we got back to the kitchen we discovered the ferret had drunk the whole quart pitcher of the double-distilled alcohol and was gone. Later we learned that he had gone to the abort—the toilet—crawled inside, and passed out. At about 6 A.M. one of the POWs found him there, stripped him of his uniform, and in the German's clothing had walked out through the main gate.

The next thing we knew the siren was going and the goons were coming in for a search, and we were marooned out on the parade ground. I had put the alcohol in nine bottles about the size of a coke bottle inside my belt, and taken it out on the parade ground. The SS set up their machine guns on the parade ground and kept us there for four or five hours, and then finally let us go back in. By that time I had buried the alcohol bottles out on the parade ground. The Germans never found the still or a damned thing.

The German ferret, who was in the clink, sang happily at the top of his voice for at least 12 hours. The poor son of a gun was really hammered.

We went back to look for the alcohol a few days later and it was gone. Someone had beaten us to it. After this incident Col. Bud Clark, a senior American officer, called me in and gave me quite a dressing down. He told me it was a terrible thing that had happened, and that the German might have gone blind or might have been shot. I don't remember that this made much of an impression on me.

(By way of background, Clark recalled that "This booze, distilled usually through an old trombone, was awful stuff. No 'temp' control ensured plenty of fusel oil and it came out the color of the inside of an old trombone—pale blue! After a number of serious incidents caused by booze among a group of pent-up, frustrated young guys, our senior American officer outlawed the distilling—much to the relief of the Jerries, who also were sure some crazy kriegie was going to get shot wandering around outside at night.")

* * * *

Charles Cook became quartermaster of South Compound. As S-4—the man in charge of supplies—Cook, among his other duties, directed the distribution to each POW once each month of a clean pillowcase and fresh set of sheets. Occasionally he managed to steal German sheets, and thus was able to provide the 10 or 15 sergeants in camp—the men who handled many of the more difficult or menial jobs in the prison—with clean sheets once a week. Cook's duties also involved breaking bad news:

> A few of our men died. When that happened I'd go to the man's barracks and get his equipment. His plate, cup, knife, fork, and spoon would have to be handed back to the Germans. I would tell the men in the building, "your roommate will not be back," and some of them would break into tears. It was really a sad place then.

Along with sorrow, there was joy over seemingly miraculous survivals in the face of almost certain death. Cook said:

> I never heard of so many remarkable stories of pilots who could find no explanation as to why they had survived. They would come to on the ground behind enemy lines, and when they came into camp they would say, "I don't know how I got down. Everyone else is dead. Why am I alive? Why am I still here?"
>
> One guy had his parachute blown off as he bailed out, close to the ground. He landed in a tree well cushioned by branches, and he got up and walked away.
>
> Colonel McNickle, later a general, hit some soft paddies in Holland in his P-47 Thunderbolt. General Stillman crashed at 300 miles an hour in his Marauder, turned over on its back, and came out walking sideways. A B-17 major, a big fat kid, couldn't get out of his plane, and rode it down. On the Ploesti raid, Rumanians killed a lot of guys on the ground with pitchforks, but didn't touch others.
>
> Our prison camp was loaded with pilots who were shot down while looking around to see if the plane they had been firing at went down. We heard that story time and time again—men shot down because they were trying to confirm a kill, rather than watching out for the enemy.

Trafficking among the compounds was strictly forbidden. Barbed wire and machine gun posts reinforced the ban, but there were some rare exceptions. The Americans had worked up a particularly good jazz band, and on occasion they were allowed to perform in the other compounds. Once, on returning from a concert in Center Compound, the musicians noted that the British in North Compound were lined up for appell, the daily roll call in the parade grounds. Accordingly, the South Compound band swung into "God Save the King." As one man, the British Kriegies snapped to attention, and remained so for the duration of the anthem. The German commandant was furious. There would be no more playing of musical instruments for one month.

The day after Christmas 1943, the Germans learned, by means of the morning appell, that four RAF officers had visited overnight in South Compound, and 13 Americans had stayed over in the North Compound.

"There was hell to pay about this," said Cook. Indeed, the rule breakers—including Barry Mahon—were promptly given extra time in "the Cooker."

* * * *

At the end of 1943, the last of the Morlaix Eagles made his shocking appearance in camp. George Sperry recalled the story of Gil Wright, starting with the explosions over Brest, September 26, 1942.

> Gil Wright bailed out northeast of Brest near the French coastline, landing in the midst of a tule marsh not far from his aircraft. He remained in the marsh in cold water up to his armpits while the Germans examined the airplane wreckage. After three days, when the Germans were no longer in the area, the French rescued him. For nine months they passed him from one underground cell to another. Finally, dressed as a priest, he was picked up by the Gestapo at a Paris railway station. For the next six months he was pretty well worked over by the SS, trying to get information out of him. When he finally showed up at Stalag Luft III, he weighed about 90 pounds and was in very poor condition.

Within Stalag Luft III, information about new escape attempts sifted through the camp grapevine. Cook recalled watching day after day as two Norwegian prisoners carried parallel bars into the open area and practiced gymnastics while tunnel-digging was in progress for what became known as the Trojan Horse escape. One of the least appealing-escape techniques was that employed by a Canadian, who buried himself with an air tube in a malodorous "honey wagon" and made it all the way to the Czech border.

On March 24, 1944, the news spread that 80 officers had escaped by tunnel from the North Compound, the largest mass breakout of the war up to that time. A recount showed that actually 76 escaped, among them Flight Lieutenant Gordon Brettell, who had led the final ill-fated mission of 133 Eagle Squadron in September 1942. Eventually 73 were recaptured. Three days after the breakout from the British compound, the prison camp commander ordered that inmates must turn out for repeated appells each day, instead of the two-a-day roll calls previously required. In South Compound, the Americans showed their rebellious spirit by refusing to stand still during the count. The insolent milling about ended the next day, however, when the Germans brought in heavily armed extra guards and stationed them behind the kriegies, ready to shoot.

Further gestures of overt resistance ended and gave way to flinty hatred with the April 7, 1944, announcement, that, on orders from Hitler, 41 of the Stalag Luft III escapees had been executed, ostensibly for resisting arrest or for trying to escape after arrest. Hitler had proposed to have the entire escape group executed, but when advised that this might bring reprisals against Germans imprisoned in other lands, he revised his execution list to "more than half" of the escaping men.

The notice of the board at South Compound on April 15, 1944, read:

> The Senior British Officer regrets to announce the deaths of the following officers, 47 in number, who escaped on March 24, 1944. This list was given to him this evening.

Listed below were 47 names, in alphabetical order, of the British, Canadian, South African, Polish, Czechoslovakian, Australian, New Zealand, Argentinian, French, Lithuanian, and Norwegian officers.

On May 19, the Germans shot three more men: two Polish fliers and a British navy pilot. Only three of the escapees made it back to England. Brettell was on a train bound for Danzig when he was recaptured, only to be shot with three of the others. "The Germans cremated the men they had shot, and then set the urns of ashes out in front of us," Jackson said. "'This is what happens to all who try to escape,' they told us."

During the remainder of 1944, the sobering impact of the failed "Great Escape" was felt throughout Stalag Luft III, including among the first caged Eagles. True, the escape artist M.E. Jackson did try one more escape, "from the West Camp, which was new and had not yet had all escape avenues blocked off." Caught on his way out, Jackson drew fourteen more days in solitary. For most of the Eagles, however, 1944 would be a year of more sedentary pursuits.

10

Caged in Two Camps

THE FORMER EAGLES' SUPPORT was crucial to bombing missions in the latter part of 1943 and throughout 1944, by which the Allies proceeded to wipe out major military targets and devastate cities in Germany. Early in 1944 the Allies started massive air operations to knock out German fighter production facilities and to destroy German aircraft on the ground and in the air, as a prelude to the Normandy invasion. Bomber missions rolled out with a vengeance into the heart of Germany.

The year 1944 also witnessed the single largest annual tally of newly caged Eagles. Stalag Luft III's population swelled, and some Eagles entered another big German POW camp, Stalag Luft I, at Barth, near the Baltic Sea, some 225 miles north and west of Sagan. Eventually, 9 of the Eagle clan were members of the Barth brigade, compared with 26 of their fellow pilots at Sagan.

On October 8, 1943, Robert G. "Pat" Patterson was flying close escort in his Thunderbolt to B-17 Flying Fortresses bombing Bremen, Germany. A veteran of 70 combat missions in Spitfires and P-47s, his awards included the Distinguished Flying Cross, the Purple Heart, and the Air Medal with three Oak Leaf Clusters. He described his final combat mission and ensuing odyssey:

> On the 4th Fighter Group B-17 escort mission to Bremen, a Messerschmitt Me 109 shot up my hood and set my plane on fire. When I finally got the canopy off, my exit was quite a fast one.
>
> As I parachuted down, I could see a great mass of flame where my P-47 had smashed into some houses. This was in the Dutch town of Onstwedde, later known as Staadskanal, 10 miles west of the German border. A long black car was speeding toward the fire, and I could hear a siren

blowing. I landed near some buildings and ran behind them as soon as I could. I did not have time to hide my parachute, but I ripped off my leather flight jacket and tossed it under some bushes. I pulled my pants legs out of my flying boots and walked out. A crowd was gathering, watching people who were trying to put out the flames.

As I stood there, a man on a bicycle pulled alongside and motioned me to get on behind him. I did, and we were off. I felt that at least for the moment I was in the hands of a friend.

The friend turned out to be a member of the Dutch underground, Aren Wiendels. The last I have heard of him since, he has been a farmer in Australia. Wiendels took me to other members of the Underground, and they hid me out and moved me several times. They said there was quite a search for me going on. This moving about continued for several days. "The Germans have your description," one of my friends told me. "They know that you are a redhead. They have been shaking down every home and every building in the area."

My underground benefactors briefed me as to what to do and where to go. They taught me enough Dutch to be able to buy tickets on trains and streetcars, which was to be my mode of travel. They gave me a false identity card with the name of Van der Heide.

One night after a British raid, and after the all clear had sounded, my friends covered my head with bandages and removed me, as a bombing victim, to the big city of Groningen, about 20 miles away. During this trip I was briefed as to our destination and our travel route.

From Groningen we moved on—sometimes on foot, at other times by bicycle, car, or train—to Leeuwarden, near the Zuider Zee. Here a Mr. and Mrs. Boersma helped hide me out and procured for me a Dutch passport with my picture on it. Later President Eisenhower gave them priority to come to the United States for the help they gave Americans. They lived at Bellflower, California, for a while, and later moved to Australia. One of my good friends, too, was a Dutch policeman known to me only as Jon, who had gone underground when the Germans invaded. Jon had nothing but nerve—great bravery. He was killed, finally, after leading the group that liberated prisoners from Leeuwarden Prison.

My Dutch friends passed me along to The Hague, and then started me to Utrecht. At night the train was stopped on a high bridge and searched. I was in a compartment with five civilians, all strangers to me. Seated next to the door, I saw the conductor and a German officer coming our way and had my ticket and passport ready. The officer looked at my passport and asked me a couple of questions. My heart was beating so hard I honestly believed they could hear it, but I kept my voice from trembling. They went on and checked the other passengers, and I felt that an eternity had passed. It took a while for my pulse to return to normal.

The train got underway again. At a station farther on, following instructions, I made a contact with a French woman, who was to see that I got into Amsterdam. This woman had escaped some way from Fresnes Prison near Paris. She was a very cool character. When we walked down the darkened streets, she carried a small pistol in her hand. When we came to a lighted area, she would drop it into an open handbag.

One day, proceeding to an address in Utrecht, she told me to catch a certain streetcar and get off at a certain street. "Get on the car and sit down so as to attract no attention," she said. "I will follow on the next car."

I caught the car as instructed. Sitting on one side was a German Kriegesmarine with a girl companion. I sat down about two feet from them, telling myself this might be the safest place. It was. His interest was centered on the girl. He paid no attention to me.

That night I left Utrecht by train for Amsterdam. My French escort traveled by automobile and was there to meet the train when I arrived.

As I stepped out of the coach, I picked her up and followed her by ten or twelve feet, as I had been told to do. We came up through the lower level to the main station on the first floor. There at the exit gate stood a ticket taker and a German officer. I thought that this time I was a goner for sure. I would have been but for my resourceful French escort. She walked right up to the German and asked for a match. As he found one for her, I stepped past, handed in my ticket, and went on out, expecting at any moment a shot in the back.

So far so good. The French girl caught up with me and walked on ahead. We finally went into a Dutch apartment; she turned me over to another escort; and we moved on. I shall always remember that courageous young woman, risking her life for just another stranger.

That same night or perhaps the next—I can no longer remember for sure—I was taken to another Amsterdam hideout and then was told to go to still another where two individuals would meet me. The meeting came off, and the two men walked me to another apartment. As we proceeded up a flight of stairs I noticed that the two men were right behind me. We went into the apartment, and a tall, heavy-set man shook hands and told me to take off my coat and hat and sit down.

Just as I put my hat down an adjoining door flew open, and four or five men rushed in. The four in civilian clothes had Schmeiser machine guns. An officer aimed a Luger pistol at me. Two more civilians who had been my so-called escorts also had pistols out. I was in the hands of the Waffen SS after being a fugitive for 53 days.

In a very short time I was up against the wall. I had no gun, and my hand had moved only part of the way to my German linoleum knife when they seized me. "English schweinhund," one shouted.

They handcuffed me from behind, searched me, and walked me down the stairs where two more uniformed men, with dogs, met us. They took me to Gestapo headquarters in Amsterdam, stripped me, and locked me in a basement cell. It was one of four cells on that side of the room, and an SS guard stood at the entrance. Every now and then he would flash a light on me.

The cell was so small I could stretch my arms and touch both sides. It was dark all the time, so I could not tell night from day. The only contents of the cell were a wooden bed and a bucket. The door was of steel. I am sure that even if I had had a tool kit, I could never have broken out.

I cannot remember how many times I was interrogated, but I do remember the first time. The guard opened the door and told me to come out, and then handed me my shoes and a two-piece white uniform. I put them on, and as I followed him out of the main cell room another guard fell in behind me. We went upstairs and into the interrogation office. I stood before a large desk, behind which was a civilian.

As he started asking questions I noticed that on the desk, directly in front of me, was a pistol. I believed they wanted me to grab for it. After a few minutes the man behind the desk picked the weapon up and put it in the desk drawer. During the interrogations, there were usually three Gestapo men in addition to the guard beside me and one in the doorway.

I was kept in my cell constantly, except for the trips upstairs for questioning. They interrogated me day and night. Now and then they handed me a small cup of ersatz coffee. Because I was almost always in darkness, I cannot say how long I was in this cell. I know that when my stay was up I was half starved, had a big bushy beard, and looked like a wild man. They came very close to cracking me—very, very close.

One day the SS lieutenant called me out for what I thought would be another interrogation. Instead he stopped and said, "You know, Patterson, you are supposed to be shot."

I told him yes, I knew that, and to go ahead; it made no difference to me. I was so weak from hunger that I did not care any more. But I was not going to inform on my friends.

"We are not going to kill you," he said. "The only reason you are still alive is that the officers upstairs respect you. They have decided to turn you over to the military. You will go to a POW camp, and that will not be too bad."

Then he asked me, "Who do you think will win the war?"

"We will," I replied.

"Yes," he said. "I agree. How long do you think it will take?"

"Less than a year," I said.

He shook his head. "I disagree," he said. "It will be more like two years."

Time proved him to be more nearly correct than I.

The guard brought me a razor and let me clean up a bit. Then he said, "We are enemies, but I would like to shake your hand." We shook hands, and I told him not to feel too badly about the way the war was going. I said that if I had been born a German, I probably would be wearing his uniform. "That is exactly what the officers upstairs said," he replied.

Two German MPs took me to Dulag Luft I in Frankfurt, Germany, for questioning by the military. For the train trip to Frankfurt they had me wear a German Luftwaffe coat. The civilian population was in a hostile mood because of the relentless heavy bombing raids on German cities. I assumed that my guards feared that in my own uniform I might be lynched by the angry mobs.

At Dulag it was more solitary, but at least I could see daylight and was given a little to eat. It was like a picnic compared with what I had as a guest of the Gestapo.

A man in civilian clothes with a Red Cross arm band came into my cell one day, saying that he was from the International Red Cross at Geneva. He started asking me questions, and I asked to see his clipboard. After looking at the questions on it, I drew a big X across the page and told him he was a German. "Yes," he shouted. "I am. And you will rot here." He stomped off.

It was now late in December 1943. I spent Christmas and New Year's Day in solitary. A couple of months later, along with hundreds of other POWs, I was transported in 40-and-8 freight cars—the type that in World

40 men and 8 horses—to Stalag Luft I at Barth, on the Baltic
　　　　weak during the train trip that I would not have escaped even

　　　　in the Barth camp, Patterson, like his fellow Eagles at
　　　　eenly concerned with escape opportunities.

　　　　e time at Barth life was uneventful—locked in at night, roll
　　　　breakouts. I had a visit from RAF Wing Commander Ferris.
　　　　of the prison compound's Escape Committee. As Big X, he
　　　　oin in escape preparations, and I became involved in three
　　　　attempts.
　　　　geant-Navigator Don Martin of the Royal New Zealand Air
　　　　arted digging an escape tunnel and got caught at it. This
　　　　in the cooler for both of us.

Another time two other Americans—a fellow from Ohio named Robertson or Robinson and another man whose name I've forgotten—and I grabbed three Germans who were repairing the prison buildings, disrobed them, and put on their uniforms and also their German Air Force hats. We got as far as the last gate, where the guard pulled his Luger out. Off we went for two weeks in the cooler this time. The German commandant called me in the next day and said the guard had recognized me. He said prisoners with a record like mine were the ones that the guards had orders to keep a close watch on.

My last effort was with Wing Commander Ferris and Lieutenant Joe Stukas, a B-24 pilot from Chicago. Ferris had been told that his parents had died. They lived on the island of Guernsey—which was occupied by the Germans—and he suspected that the Germans had killed them. He was desperate, so we tried again.

We cut a hole in the floor one night and made our way to the edge of the barbed-wire barricade. It was an ideal time; visibility was poor, due to a heavy snowstorm. We were on our backs cutting through when the dogs caught us. Since we were quite close to one of the towers, the lights turned on us. After a bit of a hassle, we were marched off to solitary again.

Donald H. Ross, like Patterson an alumnus of 121 Eagle Squadron, joined him at Barth soon after bailing out from the P-51 with which he had just shot down an Me 109 February 25, 1944. Ross had barely squeaked into the Eagles' fraternity, having signed into 121 RAF Squadron the day it changed over to the U.S. designation, No. 335 in the Eighth Air Force. The Germans captured him near Leipzig.

The third Eagle into Barth was Paul "Duke" Ellington of Tulsa, Oklahoma, another alumnus of 121 Squadron and member of the 4th Fighter Group's 335th Squadron. On March 4, 1944, the day of the first Eagle bomber escort mission to Berlin, he was somewhat surprised at having to crash-land his flak-damaged P-51 in enemy-occupied Belgium.

I had been credited with 136 missions, and I never got a scratch—only a couple of holes in the plane one time or another. I was kind of a lousy shot, only had one confirmed victory. I had washed out of the Army Air Corps cadet class—I got cut out on low-flying problems—but got into the RAF through the Clayton Knight Committee after showing my honorable discharge from the Army.

In the 335th we had changed over from the Spitfire VB to the P-47 in the spring of 1943, for combat, and that winter we re-equipped to the P-51, introducing it to combat almost immediately. The Mustang was a kindred type of aircraft to the Spitfire. Taking off on its first mission, after only 10 or 12 hours of indoctrination, felt like going home. It was kiddy-car flying.

The squadron was part of a B-17 bomber escort mission to Berlin on March 3, the day before Ellington's final run. On the approach to Kiel, the planes were called back because of adverse weather over the target. Jim Happel, commander of B Flight, encountered engine trouble. His plane could not be repaired properly for the March 4 attempt, so Ellington was assigned to the B Flight leadership.

By the time we got in as far as the Ruhr area on the March 4 repeat mission, almost 25 percent of our P-51s had had to abort because of engine trouble. The P-51 was a brand new aircraft, and they always have bugs. Over Berlin the reaction from the Germans was mild. I engaged one plane. It dived out onto the east. I didn't chase it too far. When we had broken up with our bombers after the little engagement over Berlin, I joined four other P-51s and headed west for home low on oxygen.

We flew over Antwerp. The weather was lousy, and nobody shot at us there. We cut more to the south, away from populated areas, and then turned back to the northwest and set a course for home. I was hedge-hopping slightly north of Brugge, low over the trees, when an antiaircraft slug hit the plane and took my radiator out. A second shot went through the fuselage halfway to the tail, and a third clipped the vertical fin. It was explosive, but I still had good control.

I bellied the plane in, jumped to the ground, popped a thermite bomb into the fuselage tank, and got out of the area. I walked for about three hours in the Flemish countryside, pinpointing where I was with my escape map.

Ellington determined that he was near Belgium's northern border almost into the Netherlands, close to the town of Maldegem, about 45 miles northwest of Brussels. Outside a house he saw a farm couple. The woman asked, "Are you American?" Upon his affirmative response, they invited him in and gave him some food. "It had been snowing, and they let me dry out my clothing," he said.

Two young men came in and looked me over, and asked if I had a pistol. I gave it to them, figuring I would be better off without it. They seemed pleased and left. I thought that perhaps I had just bought my ticket to Spain.

The adults went out, leaving me with only a girl of about 14 who was eager to try her English on an American. In half an hour two trucks loaded with Germans and a dog showed up. A German sergeant speaking excellent English greeted me with the standard Nazi message on occasions such as this: "For you the war is over." Goodbye Spain!

Ellington was housed for two or three days in the garrison of an anti-aircraft battery, and then was removed to a prison in Brussels. He remained there until there were enough Allied prisoners on hand to fill a railway car. Then the group was transferred to the Oberursel interrogation center at Frankfurt for "a lot of questioning."

> The British had trained us to respond only with name, rank, and serial number. After seven days of that, the Germans at the center got tired of it and shipped me to the transfer camp in Frankfurt am Main. The Eighth Air Force bombed us two times, and the British bombed us twice also. They hit installations on both sides of us, and we could feel the impact of the 1,000-pounder that the British bombed throughout the bomb shelter. From here they moved us to Stalag Luft I in Barth.

Ellington soon became reacquainted with other inmates there, including Russ Spicer, who was C.O. of another P-51 squadron, and Hank Mills, formerly of 71 Eagle Squadron, who was shot down on March 6. Summarizing their experience together, Ellington said, "We spent the time playing bridge and poker, fussing about the food, and for the whole 14 months I was there all of us were trying to get out. There was no brutal treatment. The Luftwaffe was living up to the Geneva convention to the letter, at all times."

* * * *

George Carpenter, of Oil City, Pennsylvania, a rising star among the Eagles, was on a tear in the early months of 1944. He had arrived in Britain in February 1942, reporting first to an advanced flying unit in Watton, Norfolk, and then to 57 OTU operational training unit in Hawarden, Wales. While most of the pilots were given commissions as pilot officers, Carpenter was one of the few who remained an enlisted man, with the rank of sergeant pilot.

At the end of May 1942, Carpenter was assigned to 121 Eagle Squadron at Rochford, near Southend-on-Sea. Here he was pleased to be flying with fellow Yanks, at last, and to pilot the powerful Spitfire VB with its Merlin 45 engine and its four .303 machine guns and two Hispano 20 mm cannons. In June he was flying frequent operational missions.

Three months later the RAF handed its Eagle squadrons over to the U.S. Eighth Air Force, and 121 became the 335th Squadron of the 4th Fighter Group. Carpenter finally received a commission, and by mid-1943 he was a captain.

Carpenter shot down his first enemy plane, an FW 190, in December of 1943, flying a P-47 Thunderbolt. His second air victory came a month later.

"We were just getting to the edge of Germany then," he said later. "Soon we got our belly tanks and were able to escort bombers on longer missions, deeper into Germany. Then we re-equipped with P-51 Mustangs and things really got cranked up."

The former sergeant rose to the rank of major in January 1944, and became 335th Squadron Commander. This meant he was leading every sortie, and the missions were getting hotter—more enemy planes were being shot down; more squadron pilots were being lost.

The Berlin raids of March 1944 contributed to the caging of still more Eagles. Two days after Hank Mills joined "Duke" Ellington as a POW, Sel Edner met a similar fate. Two more former Eagles, Kenneth Smith and Kenneth Peterson, became prisoners later in the month. Carpenter's score for March was six enemy planes downed. Kills by other former Eagles in that month totaled ten for Don Gentile, four each for Duane Beeson, Jim Clark, and Jim Goodson; two each for Steve Pisanos, Howard Hively, and Don Smith; and one each for Vic France, Ken Smith, and Don Blakeslee. The 4th Group's score for March 1944, 156 enemy planes, was the greatest for any USAAF group in a single month throughout the war.

On an April 5 sweep Carpenter, Beeson, Gentile, Goodson, and Vic France each shot down enemy planes, but Beeson had to bail out and became a German prisoner. Carpenter led another bomber escort mission over Brunswick April 8, knocking down two German planes. Gentile, Bud Care, and Jim Happel also scored wins, contributing to a 4th Group all-time record of 31 planes destroyed for a single mission. Former Eagle Frank Boyles was killed in the continuing air battle.

On the next mission Carpenter shot down two enemy planes and Care accounted for another. Bud Care's plane was damaged on an April 15 mission, and he had to bail out and spent the rest of the war in Stalag Luft III. Three days later, on April 18, 1944, on a bomber-support mission, five-plane ace Vic France was killed, and Carpenter was knocked out of action. His total score, under the fractional calculations then employed that took into consideration destructive shared attacks on aircraft on the ground, was 13.33 German planes. At the end of the war Carpenter still ranked fourth among all the Eagle aces, outscored only by Gentile, with 22 planes shot down and 5 destroyed on the ground, Beeson with 19.33, and Blakeslee with 15.5.

"On that big Berlin raid of April 18, an Me 109 crippled my Mustang and then I was finished off on the deck by an FW 190 in a head-on attack," Carpenter said. "Next day it was the train to Oberursel near Frankfurt, and then on to Stalag Luft III at Sagan. We were the first contingent of prisoners into the new west compound there."

* * * *

Richard Lear "Dixie" Alexander of Piper City, Illinois, had more than his fair share of adventure, on land as well as in the air. He flew 168 fighter missions in Europe and the Mediterranean in World War II and knocked down six enemy planes, two of them from RAF Spitfires and the others from USAAF Mustang P-51s. His own planes were shot up badly three times. In one of the battles he was severally injured and was given the Purple Heart. He also won the Distinguished Flying Cross, the Air Medal with 12 Oak Leaf Clusters, and assorted other decorations. Once, when his engine quit at 20,000 feet over the English Channel, he maneuvered his stricken plane 35 miles to a safe landing at the nearest English airdrome.

In 1943, flying a P-39 with North Africa as his destination, he had to land instead in neutral Portugal. When he and six other American pilots refused to board a ship that they had been ordered to take from Lisbon to London—refused because they had learned that the Germans were set to bomb it—they were thrown into prison to await trial on charges of desertion, for which the penalty could be death. Flown to Algiers for court martial-proceedings, the renegades were found guilty—"Guilty, guilty; the sound of it rang in my ears," Dixie Alexander admitted, as his fate was decided. But the sentence was a slap on the wrist: two weeks of confinement, with "confinement not to conflict with flying duties."

On May 29, 1944, in combat again, he was escorting B-24s over Austria when bullets cut his P-51's fuel line, forcing him to crash-land in a meadow. He then set out on an adventure that contained what he later claimed was the most frightening experience of his life.

> We were escorting B-24s to Wiener-Neustadt. Three Me 109s flew directly through the formation, pulled up, half-rolled, and started back at them.
>
> I gave the call to Blue section to drop belly tanks and switch on the mains, and we immediately went down after them. They made their pass from the rear and went under the bombers and down, heading for the deck with us in close pursuit.
>
> I singled out one and followed him through several spirals, waiting for him to level out and in the meantime dropping 15 degrees of flap and throttling back to keep from overshooting him.
>
> As we reached the deck, he leveled out and started up the Danube Valley. I was still throttled back, and was able to shoot from about 200 yards. Pieces flew from his aircraft in three good successive bursts. He started to smoke, half turned, and flew into the side of a hill.
>
> I started to pull up and gave the engine more throttle, whereupon it began to sputter. A quick glance confirmed that I was on main tanks. Since I was practically on the deck, I had no time to seek ready solutions, nor could I gain enough altitude to risk bailing out.
>
> I let the nose down and aimed for a couple of small clearings—little mountain pastures straight ahead, separated by a creek. I hit one pasture, bounced twice and over the creek, and came to rest in pines at the far end of the second pasture. I am sure that these two small areas, each only about 150 yards in diameter, were the only level ground in that part of Austria.
>
> Other than banging up a knee and hitting my head on the gun sight, I was not in bad shape. Immediately I got out of the plane with my maps,

escape pack and pistol, and the fire-bomb canister. Stray bullets must have nicked a fuel line; I smelled gasoline and saw four or five holes just below the cowling.

The wind was in the pine trees and it was very quiet on the ground, after the roar of the motor and the steady chatter on the radio. I was placing the canister on the wing, where it could set the plane on fire, when I heard a noise and turned.

One of the wildest-looking men I had ever seen came charging over the hill. He had copper-colored skin, dark hair, and a long mustache, and wore only a pair of tattered shorts. He was barefooted and brandished a long curved knife.

"A scimitar," I thought. "My God! I've landed in Turkey!"

I pulled my .45 from my boot and hurriedly injected a round into the chamber. The figure bearing down on me halted and started waving his hands, yelling "Ami! Ami!"

Further demonstrating friendship, he threw down the knife. In broken English he told me he was a French forced-laborer and had seen me come down. The knife he carried was for harvesting.

He pointed to smoke over the horizon, and said it was from the German plane I had shot down. Some troops were just over the hill, he said. He thrust a grubby sandwich into my hand and urged me to run.

I put the sandwich in my pocket and said, "You run." Then I pulled the pin on the firing canister and took off down a gully running into a new world. For me the war was over.

My first concern was to distance myself from my burning plane, remove myself from the area that would be the soonest to be searched. I had my small compass, but my maps were of no use because I did not know just where I was. I was on the side of a valley that seemed to run generally to the southeast. Inasmuch as this was the way to Yugoslavia, and what I considered my best choice, I set out in that direction.

Both sides of the valley had high hills with rock and timber, and they were dotted with small farms and fenced tracts of land. The sandwich that the French laborer had given me was a morale-builder—supplementing the concentrated food pills of my escape package, helping to ease the hunger pains when they started. I pushed along until dark, fording icy creeks, climbing rugged hills, and then lay down in some brush for a few hours of fitful sleep. It was very cold when I moved on again in the grey dawn.

I saw few people, and only from a distance. None of them paid any attention to me. In the afternoon of the second day of walking I came out of a wood, climbed through a barbed wire fence, and started to cross a field. Half the way to the next fence I sighted a large bull at the same time that he discovered me. He threw up his head and tail and started to trot toward me. I started running for the next fence. He increased his speed, and I increased mine until I was on a dead run. I had only a bit more than 100 yards to go, but I had had little sleep and was tired from trudging in soggy flying boots. By the time I had gone half the distance I was nearly spent. I kept on, gasping for breath, and managed to roll under the fence into a small gully. The bull came up to the fence and stood there, snorting and pawing. I was fearful that he might push the fence over. He did not.

As I lay there helpless, immobilized by fatigue, I was horrified to see another bull that could have been a twin of the first one, trotting in my direction, and on my side of the fence! I tried to crawl farther up the gully, but my legs simply would not respond.

I have never been so truly frightened in my life. I come from farm country, and I have seen what an enraged bull can do to someone that he has managed to knock down. I pulled out my .45 and injected a round into the chamber. I decided that if he came at me, I would try for a knee shot. He did not come for me. Instead, he ran up to the fence and challenged the other bull. They stood across the fence from each other, scooping dirt with their forefeet and throwing it backward. They had forgotten about me.

Slowly I managed to crawl on hands and knees down the gully until I was out of sight, and then made for the far fence and the safety of timber. Another night of sharp cold and little sleep, and I found that small paths and roads, and other signs that I was approaching a village, were appearing. An old hat was hanging from a fence. I appropriated it. With my black flying boots, all insignia removed, my large flowing moustache, and now the hat, I could reasonably pass for a forced laborer.

From a screen of bushes I observed four German soldiers, apparently guarding some laborers hoeing a large potato patch, moving away from me. Several hoes and some uniform coats and other outer apparel lay piled near me. I was tempted to steal the clothing but instead simply appropriated a hoe, and slipped back down a bank and along a stream.

Late in the afternoon, from the brow of a hill, I saw a village in the distance, and before it several squads of soldiers and some civilians and children, moving about apparently searching the countryside. Two of the soldiers were using dogs on a leash. Later I was to learn that three of our heavy bombers had gone down in the vicinity and that a hunt was on for survivors among the crews.

Fearful of having my trail picked up, I climbed into a tree and remained there, cold and uncomfortable, until searching parties and lights had all disappeared, about midnight. Then, carrying my hoe, I proceeded down a road through the dark, and by dawn I must have walked five to ten miles. I found some small stacks of hay, crawled into them, and slept for a few hours under the warming rays of the sun.

I awakened to the greeting *Guten morgen*, and saw a pretty young woman carrying two pails of milk. Quickly, I informed her that I was French and did not know German. Unperturbed, she extended a dipper, and I drank not one but three cupfuls. We walked together down the road for a short distance. When she turned off on a path I thanked her and we said goodbye.

As I approached the outskirts of a village, three small boys came down the road toward me. We spoke, and they went about 20 feet past me, then stopped and called out something in German. I pretended not to hear, and they turned and started following me, asking questions. When I did not reply they became more and more excited. Three more youngsters joined them, and soon we were a parade of a dozen or more. Looking for some form of escape, I turned off toward a cottage that had a crucifix on its gate, waved a hopeful goodbye to the children, and walked up the path and into the house.

The occupants—an elderly couple and two children—looked up in alarm as I entered. I made it plain that I wished something to eat. When the man showed signs of hostility, I motioned to the pistol in my boot. As the woman made me a ham sandwich I checked the back door, but found that the Hitler kids already had it covered. I munched the sandwich while my hosts stared at me. Since the situation appeared hopeless, I conveyed to the man my wish to be taken to the local police station. The young brats outside formed around us as soon as we came out of the house, and we all set off together.

Several boys ran ahead and soon returned with two soldiers. My guide allowed me to make a latrine stop, where I managed to dump my .45 without being seen. At the station searchers found my maps but they never did notice the compass in my boot lining. About two hours later, two Wehrmacht soldiers arrived at the station and took me to their barracks on the town's outskirts.

At the barracks a Wehrmacht first lieutenant tried to question me in halting English. I gave him only my name, rank, and serial number, and he had me locked in a cell about four by eight feet, with nothing but hard planks to sleep on, and went off for dinner.

A sergeant brought in what looked like grass soup, with two pieces of bread. He spoke a little English, and I insisted to him that as an American Oberleutnant I was entitled to bedding and better treatment. He brought me a mattress, sheets, bedding, and a pillow, and even gave me a cigarette. I went to bed at once and slept straight through until morning.

When the lieutenant returned in the morning and saw the bedding, he flew into a rage and had it removed immediately. I was glad that the sergeant who brought it was off duty.

A huge man—about six three, weighing possibly 275 pounds—in a blue serge suit, black shoes, and pale blue turtleneck sweater, came in. "I am Otto, from the Gestapo, and you are a Luftwaffe prisoner of war," he said. The lieutenant departed.

Otto offered me a cigarette, took one himself, and lit both. He said he would not force me to give more than name, rank, and serial number— "but believe me, I am capable of getting whatever information from you I want." From then on, he asked little of me and was chiefly concerned in telling of himself.

He had gone to America with his parents as a small boy. His parents operated a butcher shop in Brooklyn, and when they returned to Germany in 1930, he took over the business. He returned to Germany some years later, for a visit, became interested in the Nazi movement, and finally decided to stay in the homeland and become a party member. "After the war I shall go back to the States," he said.

"I do not think Germany can win the war," I said.

"Perhaps not, but neither will you win it," he replied. He predicted that Germany and the Allies would sign an armistice, and soon thereafter they all would be at war with Russia. He gave me the rest of his pack of cigarettes, telling me not to show them to the lieutenant. He dismissed me with the comment that I would soon be in a permanent camp, "more comfortable than this."

An English-speaking corporal in his early twenties took me to the railroad station in the town of St. Polten. "We have about an hour before the train leaves," he said. "I have a girlfriend here I want to see. You wait here in a chair in the men's room. The attendant will keep watch on you. I will bring you beer and sandwiches. Don't try to escape. In your condition you would not get far."

I did not even try to get away. I was bushed, my feet were sore, I had no maps, I had little to run with. But looking back, I know I should have tried, because the longer you are a prisoner the farther inland you go, and the more difficult it becomes to escape. The corporal came back in about 20 minutes with more beer and another sandwich. Then he was gone for a very long time. I was getting worried because people kept coming in and looking at me, and I was just about ready to take off. When he finally arrived he was in a high state of agitation, saying "Let's go; we must hurry; we've missed our train."

He led me from the washroom out through the back of the station and into the marshalling yard, across a number of tracks, to a small tender-type engine, which was used to switch cars. The engine had a caboose of sorts. He motioned me up into the cab, where a man sat ready with hand on throttle. The corporal told me this was his hometown, the engineer was an old friend, and they both formerly worked in the marshalling yards.

My guard took the seat on the fireman's side. He motioned me to sit down in the center, on a box which contained, believe it or not, German beer. The engine started rolling. My companions, having solved an immediate problem, were in high spirits. The ride lasted about 45 minutes, during which I alternately stoked coal into the tender, drank beer, and got down on the floor whenever we passed an observation post or flagman. We met loaded troop trains. I said a quiet prayer that these people knew what they were doing. They must have because we arrived safely at a siding by a very small station, and my guard and I got off.

I was taken to a camp that appeared to be a primary flying school for young German pilots and air crews. Fifteen or 20 Americans were being held there. After an initial interrogation of sorts, I was allowed to meet them, and we were able to speak freely. Most of them were bomber crews, and knew little more about what was going on than I did.

During an overnight stay here several of the German students were allowed to visit us. They asked many questions: Was the Spitfire as good as the Mustang? What was the P-38 like? Did we consider the FW 190 superior to the Me 109? The questioners all spoke English well. They were bright, alert, and young, and certainly we had one thing in common: they loved flying. And that, I believe, was about as far as their thinking went.

Talking to them gave me a strange feeling. They might easily have been young people from the United States, or anywhere else, involved in flying. And the two wild young soldiers earlier, who had taken me on my ride in the railroad engine—they were so very much like our own youth in spirit and in attitude toward the whole war.

On our trip to Frankfurt am Main the next day, I talked with a couple of the other prisoners about escape on the train, but we were closely guarded and the washrooms were barred. A tram took us on to the interrogation center at Oberursel, where the lot of us were lined up first, and then taken to solitary cells. I was forced to give up all my clothing, and again I was thoroughly searched. The searching was my third; again, they missed the small compass in my boot. It gave me a feeling of confidence, of a sort, to know that the Master Race with all of their efficiency were not infallible.

The diet that first day was bread and water. On the following morning I underwent my first interrogation, and I learned that the invasion of Europe had started the day before and that the Allies had landed in France, in several places.

I remained at the Interrogation Center for five or six days. There were veiled threats, but no harsh treatment. Different techniques were applied—the buddy-buddy approach, and other lines of questioning that I was able to resist. There were times when the questioners were very friendly, giving me coffee and an occasional cigarette, trying to warm me up.

On the morning of my last day at Oberursel, my interrogator said, "Lieutenant Alexander, why didn't you tell us you were Dixie Alexander, formerly of 133 Eagle Squadron? We did not know that you were the same Richard Alexander flying with the 2nd Squadron of the 52nd Fighter Group. We thought you must have gone back to the States."

He showed me pictures of my tent on Corsica, my shark's-tooth aircraft, and told me many things they knew about the Second fighter and its personnel. "If you had given me this information when you first came here," he said, "you could already be in a permanent camp, where things would be more comfortable for you."

As a parting shot he asked, "Did you know that your wife is no longer in the Canadian service?" I ignored that.

"Yes," he went on, "she applied for a discharge, on the grounds of nervous disorders, and she was released from the service."

I did not learn until months later that this was true. I had not even mentioned that I was married or told them anything about my family status. The thoroughness of Nazi intelligence was something to be marvelled at.

After a two-day stopover in a holding area where Alexander and his fellow prisoners were equipped with toothbrush, tooth powder, blankets, and other minor needs, they were taken on to Stalag Luft III. As the party came up to the fence, the POWs inside called out to them. Alexander heard one shout, "Hey, Dixie! What took you so long?" and recognized the voice of Charlie Cook.

Once inside, Charles greeted me with "Welcome to your new home. We've been waiting for you." Other Eagles came up and put their arms around me, shook my hand. I shall never forget their welcome. Most of them had been there for a year or two years—some for almost three years. I was to spend my next year with them.

"I know you are hungry," the ever-hospitable Cook said. "I've got something for you." He took me to his living quarters and fed me homemade potato soup and bread—to me, almost the best meal I'd ever tasted.

For the first month I lived in a large room with about 40 other kriegies. Later I shared two very small end rooms with Cook and four other prisoners. We slept in one of the rooms and sort of lived in the other. I took on the job of sports officer for the West compound—something that occupied a great deal of my inactive time. It gave me plenty to do and was certainly a blessing.

My own stay in prison camp was not marked by anything too unusual. I experienced the lack of food that others did, the long days and nights, and the inconveniences.

I made two interrupted escapes. One, at Nurnberg, Germany, was thwarted almost immediately after we cut our last wire and reached the outside. The other, during our forced march near the war's end, lasted for two days until we reached a swollen creek and were picked up by forest rangers.

The hours, days, and weeks in POW camp went slowly, but when they were all added together, it seemed that a year had passed by in a relatively short time. The daily inactivity and the repetitious lifestyle would account for this. Stalag Luft III was not a rest camp, but I experienced little hardship that I did not expect. Looking back, I regret the wasted year, but during that year there were a few plusses. I made some fine friends. I had much time to reflect and to get to know myself better, and I developed a degree of tolerance and patience that I had never had before.

* * * *

On what proved to be his final mission as a fighter pilot, Jim Goodson in mid-1944 was skimming along at the top of the heap of World War II air heros. He already had 32 enemy aircraft to his credit—17 shot down and 15 destroyed on the ground—the greatest number that would be scored by any of the Americans who had served in the Royal Air Force Eagle squadrons, and the second highest total for any member of the league-leading USAAF Fourth Fighter Group.

Goodson had tasted World War II from the very day it began, had experienced it all too intimately. He was aboard the SS *Athenia* in the Atlantic off Scotland's Hebrides Isles en route to Montreal on September 3, 1939, when it was announced over the ship's loudspeakers that Britain was at war with Germany. The 18-year-old native New Yorker had been attending a seminar in France "to perfect my language." He had majored in modern languages at the University of Toronto and boasted that "my French was better than my English, and my German was as good as my English."

"Our ship was hit just as we were getting ready for dinner—hit by the first torpedo of the war," Goodson recalled in an interview in 1983, 44 years after the event. There was an explosion and the lights went out. Panic spread among the more than 1,100 people aboard as water poured into the torpedo-damaged areas. Emergency lights came on. Goodson and other powerful swimmers plunged into flooding sections of the ship to rescue children and their mothers, and to help other persons trapped in rooms and passageways. Eventually, lifeboats were lifesavers for the majority, holding the death toll to 112.

After the lifeboats had been launched, Goodson swam to one and managed to get aboard. A Norwegian tanker bound for Latin America, the *Knute Nelson*, picked up the survivors and turned back to deliver them to the nearest neutral port, Galway in Eire. "It was 4 A.M. when we went aboard the tanker, and the newly wakened crew members didn't even know yet that war had been declared," Goodson said.

Goodson went from Galway to Belfast, and then by ferry to Glasgow. At a recruiting office there he inquired about joining the RAF, and was advised that because England lacked the resources to train the pilots it would be needing, most of this work would be done in Canada under the Empire Training program headed by Billy Bishop, Canada's first ace of World War I. Since Goodson had already paid for his return passage to Montreal, the logical thing for him to do would be to return to Canada and sign up there with the RAF.

Goodson and other *Athenia* survivors returned to Canada safely on *The Duchess of Athol*. He progressed through primary training on the fleet biplane and through advanced training on the AT-6 Harvard, and then was sent back to England for further training in the Miles Master. He was posted first to RAF 43 Squadron, at Tangmere, and then to 416 Canadian Squadron.

Following the decimation of the ranks resulting from the disastrous Morlaix mission, the Eighth Air Force Fighter Command put out an emergency call for experienced pilots to rebuild 133 Eagle Squadron, soon to become the U.S. 336th Squadron. Jim Goodson found himself transformed, at this late date, into an Eagle.

With the absorption of RAF 133 Squadron into USAAF's 336th, another leading Eagle ace, Don Blakeslee, took command of the unit. Through bold and often harsh leadership, he rebuilt squadron morale and placed his former Eagle flying mates on a winning course. The 336th changed over from Spitfires to new long-range P-47 Thunderbolts capable of out-diving as well as out-distancing the enemy.

Blakeslee's brilliance in fighter supervision and combat leadership ultimately won him promotion to the command of the 4th Fighter Group. On August 16, 1943, however, his boldness of attack almost brought him to grief. As he dived after a Focke-Wulf 190, he found himself in the midst of a swarm of enemy fighters.

Jim Goodson, covering his leader, called out warnings and shot down two of the planes pursuing Blakeslee and possibly destroyed a third. Out of ammunition, Goodson nonetheless frightened off other enemy planes trying to attack Blakeslee, now blinded by oil streaking over his windscreen. Goodson guided Blakeslee all the way back to England and down to a safe landing at Manston. For his heroic achievement, Goodson received the U.S. Army's third highest award for bravery, the Silver Star.

In December 1943, Fighter Command transferred Blakeslee temporarily from the 4th Group command to the 354th Group of the Ninth Air Force to monitor the introduction into combat of the swift and powerful new North American P-51 Mustang. Blakeslee became so greatly impressed with the Mustang's superiority over other aircraft at all speeds and altitudes that he induced Maj. Gen. William Kepner, chief of XVIII Fighter Command, to order the plane for the 4th Group. The general ruled, however, that with a major bomber offensive already underway the 4th could not be spared from combat for more than 24 hours. Blakeslee promised that the pilots and ground crews would have the new planes ready for operation within that time frame. Sure enough, within days of delivery the Mustangs were in combat service. Goodson said this meant that the pilots had only about half an hour, instead of a couple of hundred hours normally allowed for new fighter-plane indoctrination, to get ready for the first Mustang combat missions.

On March 6, 1944, Goodson, now a major, leading the 336th Squadron, attacked a force of more than eight enemy fighters while escorting bombers to Berlin, even though 20 to 30 German aircraft were providing their fellows with top cover. He shot one German plane down and damaged another. Fragments of the damaged enemy plane's fuselage smashed into Goodson's Mustang but failed to knock out the engine. This encounter won Goodson the Distinguished Service Cross.

The scale of the offensive against Germany steadily intensified. On June 20, 1944, having flown every mission for the past four or five weeks, Goodson was in need of a rest. The mission at hand promised to be relatively routine, a mere escort of bombers attacking the Baltic port of Stettin. But at the last moment before takeoff, with all propellers turning, Capt. Don Emerson, selected to lead the squadron because of seniority, called for a five-minute delay to calm his fear that he did not feel in top condition.

Goodson quickly reassigned Emerson to fill an empty slot in one of the squadron's sections and jumped into his own plane to lead the formation. The flight turned out to be routine. The squadron accompanied the bombers to Stettin, saw them deliver their explosive loads, and observed relieving fighter escorts come in to safeguard the return to base. Goodson, sighting German fighters over the Neubrandenburg area, led his squadron there and managed to shoot down two more enemy planes.

Suddenly, he observed a tiny, stub-wing plane half-hidden on the far side of the airfield. He had discovered Nazi Germany's newest and potentially most dangerous air weapon, the Messerschmitt 163, the swiftest aircraft that had yet been developed. Goodson had been especially alert to the possibility, as he knew that the return route from Stettin would take the squadron over the Peenemunde sector, where German scientists had been conducting rocket research. Turning his gunfire on the prototype Me 163, Goodson charged directly through menacingly dense antiaircraft fire in crossing the Neubrandenburg air field at a perilously vulnerable medium-low altitude.

Exploding flak cut through his plane and slashed into his right knee. Having set his Mustang down in a woodsy field several miles from the air strip, Goodson cut the switch, pulled back the canopy, removed his seat belts and parachute, and hobbled to the ground. Rather than taking the time to try to destroy his plane himself, he motioned to the squadron circling above to do the dirty work and hastened away as quickly as he could. The squadron demolished the crippled Mustang and then headed back to base, helpless to be of further service to their unlucky C.O.

Goodson had a map and a miniature compass, and calculated that his best hope of evading capture was to make for Rostock, on the Baltic some 80 miles distant, and then hope somehow to get aboard a ship that would take him the 70 additional miles to Sweden. He also carried a first aid kit, which he hoped could at least alleviate the pain in his torn knee and shrapnel-shredded legs.

Slowly he limped his way through woods and fields. At one point, as he crouched down in a wheatfield, he glimpsed a cyclist moving along a roadway. The man alternately was singing "God Save the King" and calling out in English an offer to help. Goodson, fearful of a trap, kept silent, although much later he reflected that there might even have been some kind of an anti-Nazi underground in Germany.

Painfully, he pushed on, hour after hour. Several times he narrowly avoided encounters. Once, a passing automobile flashed lights continuously alongside the roadway but somehow missed him as he lay in the darkness. The next evening a farmer entered the barn in which Goodson was trying to milk a tethered goat. The flier scurried past the startled German and out into the darkness. Soon thereafter Goodson heard dogs baying, apparently following his trail. Desperate, he took to a tree. The dogs and their masters came close, but then the animals apparently lost the scent and they all went away.

On the third evening, Goodson swam a river and on the far side walked into the hands of two German soldiers almost before discovering their presence. They took him to a village jail, where he was questioned in German. Goodson answered the queries in German, professing that he was a French worker who had been injured by bombing airplanes, had become separated from his labor party, and had lost his identification papers.

Later, after an hour-long truck ride, he was placed in a jail cell and was questioned much further. He was told that, with his fluency in both French and German, he was on the face of it a spy and saboteur. Goodson related his experience to friends later, saying:

> I had been told by one of our intelligence officers that in the event of my capture I should say I was a French worker. This proved to be unwise.
>
> The Germans were very worried about their slave laborers. They had many more prisoners than men to look after them, and they were afraid of uprisings. They said my name was in the Gestapo file in the nearby town of Demmin. I was a spy and saboteur and would be shot.
>
> I spent the night thinking that I would be shot in the morning. I finally managed to see the commanding officer and talk to him. I told him I was not a French worker and not a spy, but a very important Air Force officer. "Reichmarschall Goring will be upset if you shoot me," I said. "All you have to do is query the Luftwaffe."
>
> I knew that the Luftwaffe and the Gestapo did not like or trust one another. "You will not be making a fool of yourself if you tell them about me," I told him. "You can make a fool of them. Tell them you have learned

that their organization has been looking for Goodson for two weeks and it is about time someone did something to capture me."

Warming somewhat to his prisoner, the German offered Goodson French cognac and then a Cuban cigar. Goodson blew a series of smoke rings with the cigar. The German smiled and said, "You are not a French worker. They don't smoke cigars."

The German picked up a telephone and spoke briefly. Then as he was about to leave, he told Goodson the Luftwaffe would pick him up within an hour.

Writing of this in his sensitively-done and beautifully descriptive book, *Tumult In The Clouds* in 1983, Goodson said,

> I could not believe it! I dared not believe it! It seemed too miraculous to be true that I was to be allowed to live after all. It was like being born again. Everything seemed new and wonderful, and I saw it all in a new, clear light."

A Luftwaffe escort was assigned to take Goodson by train to an interrogation center at Oberursel, near Frankfurt. They would have to change trains in Berlin at a central station, Friedrichstrasse Bahnhof. They arrived at the station during the noon hour just as air raid sirens were wailing, and they crowded with other passengers into a large shelter underground. Exploding bombs shook the structure.

"It was highly unusual for me to be on the receiving end of a mission that I had helped to plan," Goodson recalled four decades later. "I knew that a raid would take place at that time, on this date, and I knew that the aiming point would be Friedrichstrasse station.

"The 'all clear' sounded, and people started to leave the bunker. I called out that the raid was not over yet—another wave would be coming in. My Luftwaffe lieutenant repeated my warning to the crowd, and just then the warning sirens came on again. Everyone moved back into the bunker."

In his book, Goodson said that after the raid, outside the bunker, it was "a hell on earth, a maelstrom of ruins, blazing buildings, crashing walls—and everywhere bodies; writhing bodies, maimed bodies, dead bodies . . . the London blitz all over again, but a hundred times worse."

"It was a holocaust," Goodson said in an interview long afterward. "My guard asked me to parole myself in order to help with the rescue work, and to free him of the responsibility of watching me. At first I refused because it is the duty of prisoners of war to try to escape. But then I looked around again at the disaster and at the bleeding, suffering people, and I gave the escort officer my word that I would not run away. With that, they removed my handcuffs."

Observing that there were few able-bodied young men such as himself around to assist, Goodson plunged into the rescue-crew task of pulling the dead and the dying, the wounded and the shellshocked, from the debris and from still-smoking ruins. When most of the victims had been removed, he burrowed through debris from which screams could be heard, and recovered from under the body of its mother a sobbing child about two years old. The little girl stopped crying and clung to him tightly. Emerging from the hole he had dug out, Goodson at first was so dazed and confused that he refused to give the baby over to a Red Cross woman, insisting that he had promised the mother as she lay dying that he would take care of her child forever.

At Auswertestelle West, the Luftwaffe interrogation center at Oberursel, Goodson was brought before the chief interrogation officer, Hanns Scharff, an affable man perhaps 10 or 15 years older than the prisoner.

> He welcomed me, shook my hand, called me Goody, a nickname that also had been used as my call sign on some missions. He had my photograph up on the wall behind his desk, along with those of Blakeslee, Gentile, and others. In excellent English he told me it was too bad I could not go back to my base at Debden because last night there had been a big party at the Rose and Crown—Bubbles the barmaid was there—to celebrate the fact that the 4th Group now had destroyed 1,000 enemy planes.
>
> I gave my name, rank, and serial number but was surprised that he did not press me for more information. He already knew it all. "There is nothing you can tell me, Goody," he said. He added that the Pentagon had notified my mother that I was missing and presumed dead. The German government wanted to let her know that I was all right.
>
> When I declined to give him my mother's address, he said that while I probably did not know it yet, my mother had gone to the Bahamas, to Nassau, for a rest. She would be contacted there through the Swedish Red Cross. Sure enough, I heard from my mother a few weeks later that she had received the message, her first news that I was alive.

At Oberursel, Goodson received kindly treatment, decent food, and badly needed hospital attention for repair for his fractured kneecap and removal of flak particles from his legs and buttocks. Then he, too, was moved to Stalag Luft III.

* * * *

July 13, 1944, was a Friday—a traditional bad luck day. For Major Wilson V. "Bill" Edwards, formerly of the Royal Air Force 133 Eagle Squadron, and now Director of Operations of the 4th Fighter Group, U.S. Eighth Air Force, good fortune eventually outweighed the bad—though barely.

For five successive days and nights, fleets of Allied heavy bombers, escorted by P-51s and P-38s, had been attacking targets in or near Munich. On this day Edwards was leading the 336th Squadron on his third sortie to the Bavarian capital.

> I was sitting right on top, working with the bombers somewhere near Stuttgart, Germany. At 26,000 feet, flak knocked out all my instruments and

set the plane on fire. I headed toward France and rode the plane down until I was on top of the undercast.

I grabbed my maps and navigation charts and rubbed out the lines that would show the enemy our course. I got rid of the canopy and went out the wrong side, because of the plane's torque, and I remember hitting the tail. I thought I must have broken my neck, or my legs or back. I tumbled down through the clouds, upside down at first, and I pulled the muscles in my shoulders trying to right myself. The countryside below looked so wild and uninhabited that an invasion force could have landed there and not been discovered for a year. And I landed, of all places, in the backyard of a constable's home.

The man didn't even let me get out of my chute. He started jabbing at me with a pitchfork. Other people gathered around, hitting me. The plane must have smashed into a barn near the village. There had been a hell of an explosion, and I could see smoke.

Even though I felt like I had a lot of broken bones, they made me strip—take off everything—and then they started clubbing me. Women and children were there, too. I would bend over from a shovel blow in my stomach, and then someone behind would straighten me up with a pitchfork stab in my rear.

They took me, still naked, to a sort of town hall and put me in the middle of a room, while a crowd was gathering. Then they opened two doors so that the people could come in one, walk past and spit on me, and then go out the other door.

Finally the Wehrmacht—the Army—picked me up and took me to a jail on the French side of the border, at a place called Baummholder. They threw me into the jail with their own prisoners—prisoners from the desert.

Then I was with the Gestapo for a day or two. They were curious about the clothing I was wearing. I was not wearing a flying suit but had on regulation officer clothes—green pants, brown shirt, sweater and A-2 jacket, and officer's high-top shoes. Also they wanted to know about my hair. After one of the recent successful missions against Leipzig, in celebration we had our hair cropped really close, and the sideburns on one side clipped and shaved off. And I was wearing a tie. "You look like a forest meister, one of the men who take care of our woods," I was told. "Where is your pilot's shirt? You are a spy."

They tried hard to make me admit I was spying, but then something happened. Whatever it was, they moved me to Oberursel, a Luftwaffe intelligence center near Frankfurt/Main, where the head interrogator, one Hanns Scharff, fluent in English, checked me over for about a month. He promptly questioned my identity.

"Why do you say you are a major when you are really a lieutenant colonel?" he asked."You are Colonel Curly Edwards of the 352nd Group, and we have 35 pictures to prove it. We have no record of a Major Wilson Edwards."

Scharff kept showing pictures of Curly Edwards; finally, he agreed that the two men named Edwards did not really look alike. It was harder for him to accept Bill Edwards' explanation for the lack of a record from 133 Squadron—that the records had been expunged when he left after contracting spinal meningitis in the spring of 1942. In the end, however, his story somewhat begrudgingly acceded to, Edwards was processed out.

From Oberursel it was a long train ride north for the Allied flying personnel bound for Stalag Luft I at Barth. Edwards was told that, as the senior officer in the group, he would be responsible for anyone who tried to escape.

> The train halted at a station because bombers were approaching. They made us POWs stay in the cars and they took our shoes away. All the German guards and train crewmembers ran for shelter. Luckily, no bombs fell near us.
>
> One of my fellow prisoners, a B-17 crewmember named J. J. McNamara, had been badly burned when his plane was shot down. I carried him whenever we had to move, and I got to know him quite well. Later, at Barth, he was one of those repatriated home. He got in touch with my sister, in Washington, D.C., and that was the first time my family knew that I was all right.

Like Charles Cook at Sagan, Edwards was appointed quartermaster for his camp. "I had charge of distributing what clothing we had," he said. "We had a normal prison camp life at Barth. One doctor was available in camp. He was a British Army Major dentist who had been captured on Dunkirk. But there was no treatment for injuries such as mine—cracked vertebrae, a herniated disk, both legs damaged."

* * * *

On August 3, 1944, Major Fonzo Don "Snuffy" Smith, a veteran of 121 Eagle Squadron, embarked on his 113th combat mission, escorting B-17 bombers. He was flying a new P-51 Mustang fighter with a Packard-built Rolls Royce engine that had only 35 hours of flight time. Credited with having destroyed three enemy planes, Smith had been awarded the Distinguished Flying Cross and the Air Medal and assorted Oak Leaf clusters.

Smith had recently become group operations officer following a chain of events dating back to his 30-day leave in January, 1944, when he went home to Paris, Texas.

> While I was in Paris—the Texas Paris—Colonel Tommy Hitchcock, the famous polo player I had known in England when he was air attaché at the U.S. Embassy in London, sent a plane down to bring me to Abilene, Texas, where he was putting together a new Group to return to the European Theater of Operations.
>
> He asked if I would like to go with his group as Operations Officer. I said I would. However, before that could be arranged, Washington ordered that no more Groups would be sent overseas. From that time on, pilots would be sent as replacements only. When I got back to England, I called Don Blakeslee at Debden, and he sent a plane for me. The day that I arrived, the Group Operations Officer was lost on a mission, so I inherited the job.

Smith said he had time in that new post only to fly a couple of uneventful fighter sorties, and then came August 3, 1944, and the ill-fated bomber escort mission that "looked like it would be a milk run."

It was a beautiful day—one day after General Patton had broken out of the pocket at Saint-Lo on the Cherbourg Peninsula. As we turned for home after reaching our destination, my P-51 threw a rod through the coolant system. I knew I didn't have much time, so I headed for the Cherbourg Peninsula in hopes that I could land, or bail out, behind our own lines.

The engine heated rapidly. I called my wingman and said I might have to bail out if things did not get better. Then the engine became so hot it started to seize up, and got so rough I thought it would shake the wings off. Five more minutes would have put me behind our lines, but I had no such luck. The engine caught fire. I told my wingman I was going to bail out. I jettisoned the canopy, unfastened my shoulder and seat belts, snapped the plane over onto its back, and kicked the stick forward. I went out of the plane like a bullet.

This all happened at about 15,000 feet. The next time I saw the plane we must have been a mile apart. I waited and opened the chute at what I thought was about 8,000 feet. It was quite a jolt.

I could see a grass field below. Birds were singing, and it sounded like the most peaceful place on earth. Then I saw two soldiers running to grab me. They were waiting for me when I hit the ground, and they turned out to be from the Ruhr Valley of Germany, where their homes and families had been bombed out. They were very unhappy, wanted to blame me for all of it, called me a *terror flieger*.

They took my cigarettes, my lighter, and my watch. Luckily, I didn't have my gun. I didn't want to carry a gun on a mission of this type because all it could do was get you killed. They were getting ready to work me over with their rifle butts when a German Army lieutenant came up.

Our British trainers had told us that if we ever became prisoners, we should remember that we were officers and demand to be treated as such. "Don't ask; demand," they said. I told the German lieutenant, "I am an officer and I insist upon being treated as one. These men have taken my belongings. I want them returned."

The lieutenant ordered them to hand back my property, and they did. I could tell by their looks and attitude, however, that they felt like taking him on. They went away grudgingly.

The lieutenant spoke to me in English, but that did not surprise me. I never met a German who did not speak English. Even cadets that I had known, if you made a grammatical error, they would catch you up on it.

When I told him I was from Paris, Texas, the lieutenant said he had formerly lived near San Antonio at New Braunfels, Texas, which had been founded almost 100 years earlier as a German colony. I got the impression that he knew more about Texas than I did.

He took me to a palatial French country home where German officers were staying. There were carpets on the floor. They were set up for light housekeeping, and a French woman was there cooking for them. When we went in to dinner, they pulled a chair out for me at the head of the table, and when the food was brought in, I was the first to be served. As a major, I was the highest ranking person present. Everything went according to rank. They treated me as one of their own.

That night when it was bed time, they gave me a private room with bath, and they permitted me to soak in the sunken tub, as though I was living like a Vanderbilt. But they took every stitch of my clothing and kept it all until morning. If I was going to attempt an escape, I would have to do it in my birthday suit. I had a good night's sleep.

The next morning, Smith said, things started changing—for the worse. Two men in civilian clothes arrived—"the Gestapo, looking like Chicago gangsters of the Al Capone 1930's movie era, in an old touring car with driver and outrider."

They put me in the back seat between them, with a squirt gun in my ribs. They told me that if an airplane attacked us, I could get out of the car with them, but if I got more than 10 feet away I would be shot. This I did not question at all, but what really frightened me was that it was a beautiful day with not a cloud in the sky, and I knew that all Allied aircraft had orders to shoot everything that moved on the ground in northern France.

The French had turned all the signs at intersections every which way. The Germans had road maps, but nothing matched up. We drove around in circles all day and fortunately did not see one plane. In the evening we wound up at a POW reception center that had been set up in a racetrack training stable near the city of Falaise, which is about 120 miles west of Paris. I spent the night in a horse barn.

From there we were put in a truck with a wood-burning engine and taken to a small town about 20 miles from Paris, and housed in a huge, beautiful church with 10 or 20 other POWs—British, American, and two Australians who had been with me on a ship crossing the Atlantic to England. We heard some planes dogfighting overhead and saw a P-47 get shot down.

One of the fellows said, "The Germans will be throwing that P-47 pilot in through the front door in an hour or so. Let's hit him up and see if he has any cigarettes left." When the pilot was brought in, all shaken up, the POW introduced himself and said, "By the way, you don't have any cigarettes, do you?" The pilot had a few choice words for us.

Next we were transported to Paris and put on the third floor of the North railway station. Once, during our three days there, I climbed out of a rest room window onto a ledge, with my back to the wall, and walked halfway around the building trying to get on the roof or in the attic. The Germans were scurrying around preparing to evacuate Paris, and I wanted to stay up there until we were liberated. I could not find an opening anywhere, and had to walk around and climb back through the window again.

There seemed to be panic everywhere when they put us on a train heading for Frankfurt, Germany. We really didn't think they could get us out of Paris, but somehow they did. On the first day of the trip, on the plains of eastern France, the train broke down. The Germans came around with pieces of paper, asking us to sign a pledge that we would not try to escape.

I refused to sign, and as the senior officer I ordered that no other POW sign it. Some P-38s came along and strafed the stalled train. Our fighter pilots at this stage of the war had instructions to shoot only the engines of trains—some of the trains had Red Cross markings—but in this instance they strafed the whole train, everything except the engine. By some miracle, they did not hit any POWs or Germans. Many of the Germans jumped off the train into ditches, for cover. One POW, a West Point captain, had a slight heart attack and fell over in a faint. We tried to get him a stretcher. The Germans got excited and shot a tommy gun through the car, but no one was hit.

With the engine finally repaired, the train got on the way to Germany again. Smith said that, during the second night out of Paris, he and a Scotsman, whose name he had forgotten, managed to open a window, although they did not know how fast the train was going or what obstacles lay along the track. "I got him out the window and gave him a good shove so that he would not wind up under the train wheels," Smith said. "Just as I got halfway out of the window a guard came along and grabbed me and pulled me back into the train. I have wondered a thousand times if the Scot escaped or was killed."

Smith also reported that, in order to prevent his watch from being stolen off his wrist by the avaricious Germans on the air-conditioned train, he took to wearing it in the crotch of his underwear. He added that the technique worked.

Upon arrival in Frankfurt, all of the pilots and air crew members among the prisoners were sent on to the interrogation center at Oberursel, 10 miles outside the city. Smith remembered spending 14 days there in solitary confinement in a room about eight by ten feet in size, with one small cot-type bed as the only furniture. There was a window with glass that he could not see through, and beneath it a steam radiator.

> About an hour after going into solitary, a Luftwaffe officer came and asked for details as to my mission. As I had been trained, I gave him my name, rank, and serial number and nothing else. My interrogators did not try to coax any other information out of me, but they let me keep my room for two weeks. The only people I would see were the ones that delivered my cup of barley soup. Every two or three days an officer would appear and ask if I was ready to talk. I answered in the negative and they would go away.
>
> It was August and very warm with no doors or windows open, and then about 4 P.M. they would turn on the radiator. There was no reading matter. I would take off all of my clothes except my shorts, and I slept for most of the 14 days—and lost weight.
>
> At the end of this confinement, they took me before a Luftwaffe major, who proceeded to read to me my history with the RAF and the U.S. Army Air Corps. I had assumed that they did not know who I was or where I had been captured because I had heard that the people who captured me had been captured or killed in the action back in France. I gathered that they had finally identified me by piecing together information from other POWs of our group who had passed through ahead of me.

They told me my former roommate, Kenneth G. Smith, was at a nearby hospital for skin grafts on his face. He had been badly burned when he crashed with his plane. The Germans invited me to go visit him.

I found K.G. to be in good spirits. He had been burned around his eyes when the rubber around the lenses of his American-made goggles caught fire. A German doctor had been doing a skin graft to relieve pressure on his eyelids, and K.G. said the grafts had already been very successful.

At the hospital I was surprised also to find Lt. Col. Francis Gabreski, who was not an Eagle but was a leading ace. I knew he had been shot down a few days before I had to bail out in France. He seemed to be in good shape and apparently had not been hurt.

The young German doctor, not yet 30, was quite a guy. He had been treating a British air commodore who had parachuted into France a year earlier to work with the Underground, and had had an arm torn out of its socket and turned around his back. The day I talked with him the arm had been taken out of its cast and he could use it.

The doctor also was working with two bomber crew members who had been burned head to toe. The door was open and I could hear the men screaming. "Aren't you treating those fellows a little rough?" I asked the doctor. "That's right," he replied. "I intend to make them mad and that makes the blood flow so they will get well. If they could get up they would kill me. Without this kind of treatment they will die."

After a three-day visit with K.G. Smith and Gabreski, Don Smith was sent on in mid-September to Stalag Luft III, at Sagan.

I was greeted at the main gate by men who had been my friends back in England and who had heard by the grapevine that I was on my way. Any time you have movement of people, like the 50 or so POWs on the train coming from France to Germany, you have a grapevine working. I was one of the last of the Eagle group that transferred from the RAF to the U.S. Air Force to go down. They wanted to know what took me so long to get there.

I was assigned a room in Block 38-A, Center Compound, the famous compound from which many British prisoners tunneled out only to be recaptured, lined up, and shot. After the mass escape, the British were transferred to another part of the prison, and Center Compound was made an all-American camp. I was put into a room with four other prisoners, one of whom was Captain M.E. Jackson, knocked down on the Morlaix mission.

Smith recalled the special sadness of his close friend Frank Fink.

Fink was my wingman when he went down in September 1943. I almost had a nervous breakdown when the phone rang in the officers' mess that night and a voice said, "Lt. Fink is all right. The French Underground picked him up. They (Vichy France—at Toulon) scuttled part of the French fleet, but a submarine got out."

The mother and sister of the submarine commander took care of Frank, but someone turned them in and Frank and the two women were put in a German jail. The Germans never did beat him up, but every day they beat the women, trying to force them to talk. That wrecked him for the rest of his life.

When the Germans put Frank in Stalag Luft III on September 9, 1943, the two women who had sheltered him were still in the jail. Frank was brokenhearted. Before the capture he was always a prankster. Afterward he was always as solemn as the day is long.

My POW life at Sagan was pretty much routine at this stage of the war. We had orders by this time not to try to escape because the end of the war was in sight and the risk was not worth the effort. So we kriegies mostly were playing cat-and-mouse games with the Germans.

* * * *

Inside Stalag Luft III, the lessened preoccupation with escaping gave time to undertake other, less risky activities, such as writing and passing gossip. George Sperry's account is illustrative:

> During 1944 I managed to write a few pages of poetry. I made drawings, jotted down notes about the Dear John letters that men received from girls back home they had hoped to marry but who wrote now of changes in their plans. Sometimes the photos of the girls would be posted on the camp bulletin boards alongside the Dear John letters.
> Others posted letters from family: from Mother, "Be sure and call the doctor if you feel ill, and don't try to escape." From a wife who had been asked by her POW husband to send chocolates and sugar, "Don't you know there's a war on?" From another wife, "The $3,000 you left in our joint account is all gone. You'd better increase my allotment." From a girlfriend, "I've met the nicest flying instructor. He's trying so hard to get overseas but they just won't let him go." From an old friend doing his bit back home, "Sure glad you put my name on one of those bombs you dropped on Germany. Sure gives me a wonderful feeling, knowing I'm really in there pitching."
> One Kriegie wrote a letter of thanks to a woman whose name and address were pinned to some socks in a Red Cross parcel handed out in camp. He posted her reply, "Glad you liked the socks but I really intended them for fighting men."

The prisoners in Stalag Luft III were supposed to get all of their news from OKW—*Official Kommunique Wehrmacht*—the English-language propaganda weekly distributed by the Germans. In South Compound they preferred their own five-column typewritten newspaper, *Circuit*, with its bits of information obtained from newly-arrived prisoners and from letters from home. Charles Cook figured in a story from the October 23, 1944, issue.

> ED BAXLEY'S COLLECTION
> OF "EX'S" MOUNTS
> WITH OCTOBER'S BIG MAIL
> Absence Doesn't Always Make
> The Heart Grow Fonder

Gentlemen may prefer blonds, but it's the brunettes who are putting the prod to our dashing young airmen at South Camp.

On the wall above Ed Baxley's sack, pasted row upon row, are snapshots, colored portraits and even a picture of a bride in her wedding gown.

Brunettes predominate in Ed's gallery of "ex's," photos of gals whose ardors have cooled while sweating our boys out.

He started this collection of impatient maidens when his roommate C. W. Cook received word from his "fiancee's" father that she had upped and married another. Ed got her photo.

Other Kriegies who were jilted soon beat a path to Baxley's door and the collection grew. October's mail brought more "Dear Lootenant" letters usually written by some well-meaning relative. And so the gallery was increased by several new faces.

The latest activity of Bill Hall was also featured in the *Circuit* of November 4, 1944:

"KRIEGIE" HALL
OPENS BARBER SHOP

Bill "Kriegie" Hall of the original Eagle Squadron has been in Sagan so long he decided to go into business.

This week Stalag Luft III's oldest American POW, down since July 2, 1941, opened a barber shop in the east end of the Shower Bldg., where he'll clip the boys from ten in the morning till noon and from one to four in the afternoon.

"Kriegie," who built his own barber chair, says he'll install another one if business warrants.

"For something to do, I started a barber shop and ran it until the camp was evacuated, Hall said." I was giving the guys haircuts, charging them 75 cents a haircut. They'd sign a book and leave their address, and I was to bill them after the war.

"I gave haircuts there for some time and it was lots of fun, and anyway it was a place for the fellows to sit and talk. The account book kept growing and growing, and on the night of the evacuation [January 27, 1945] it was too heavy to carry. It went into the fire."

* * * *

Don Smith was the last Eagle captured in 1944, and the last one to enter Stalag Luft III. In early 1945, the last two Eagles to be caged were imprisoned at Stalag Luft I in Barth.

Jim Gray's first operational mission as a member of 71 Eagle Squadron had been on January 6, 1942, just two days after his birthday. He shot down his first enemy aircraft, a Focke Wulf 190, on August 1, 1942. Less than two months later, when the RAF disbanded its three Eagle squadrons and transferred most of the American pilots to the U.S. command, some of the Eagles, including Gray, elected to remain with the RAF. "I liked the treatment I got in the RAF as compared with that promised by the U.S. Army Air Corps," he said.

Gray joined RAF 131 Squadron at Tangmere in September 1942, and the following January he shipped out to North Africa and joined 93 Squadron in Tunisia. From there, he proceeded to Malta to cover the invasion of Sicily.

Gray's service in the Italian and North African theaters continued through his birthday of January 4, 1945.

> I served as a flight commander in 111 and 72 squadrons, covering the Salerno bridgehead southeast of Naples and the Anzio bridgehead south of Rome. During the summer of 1944, I was assigned to No. 71 OTU in Egypt, as instructor on fighters.
>
> In December 1944 I rejoined my old 72 Squadron in 324 Wing in northern Italy. I had served with this wing from March 1943 until April 1944 in North Africa, Sicily, and Italy. At the time I returned to the wing, they were equipped with Spitfire IXs, dive-bombing and providing close support to the forward troops.
>
> I was shot down near Bologna on my birthday. My final sortie with the RAF was an armed reconnaissance about 20 miles behind the front lines. We spotted enemy transport drawn up by a bridge and proceeded to bomb (and miss). Then we went down to strafe, and I was hit in the coolant radiator.
>
> The engine started to smoke, so I said goodbye to my kite at about 1,000 feet. I was picked up almost immediately by an Italian patrol. They turned me over to a German army captain, and we drove back to the flak post which had shot me down. I was introduced all around and congratulated them on their marksmanship.
>
> On the trek to Stalag Luft, I was accompanied by a couple of feldwebels. We went by trolley, truck and train, hitch-hiking at times, through the Brenner Pass to Munich and then on to Frankfurt. One of the feldwebels had a trunkload full of goodies which he was taking back to his family, and asked me to assist with one end of it. Initially I refused, proclaiming my officer status. He got very angry with me.
>
> Arriving at Dulag Luft, I underwent the usual five days in solitary and grilling at Auswertestelle West. Then it was on to Barth, where I met up with many of my erstwhile squadron mates.

Six weeks after Jim Gray's bail-out over northern Italy and his subsequent delivery into German hands, his former Eagle Squadron colleague, Roy W. Evans, experienced the last Eagle parachute descent into the clutches of the Germans.

Rough going was no stranger to Evans, a fighter pilot from San Bernardino, California. He had joined 121 Squadron at North Weald on the last day of 1941. On the following March 9, returning from a minesweeper escort patrol, his Spitfire engine stalled and he overshot the runway. The plane was badly damaged, and Evans received a fractured arm and broken jaw.

In September 1942, the Eagle Squadrons transferred to the U.S. Eighth Air Force, and 121 became the 335th Squadron of the 4th Fighter Group, flying

Thunderbolts. Later that fall Evans shot down an enemy plane with his P-47 but had to bail out into the English Channel. He was rescued promptly.

A skillful and daring combat pilot, Evans went on to become commanding officer of the 335th Squadron and an ace, blasting six enemy planes out of the sky. He was C.O. of the 359th Fighter Group when he was shot down February 14, 1945, the last Eagle to become a POW.

George Sperry, Secretary of the Eagle Squadron Association, sketched the circumstances in the *ESA Newsletter* of July 1983:

> In the closing months of the war, while leading his Group on an escort mission over Germany, Roy was again shot up and forced to bail out, this time breaking both legs when he ran into the stabilizer for the third time. When he hit the ground he was surrounded by a large group of angry German women, who were about to run him through with pitchforks, but was rescued by a platoon of Gestapo SS troops that saved his skin. He spent the rest of the war in a hospital bed and then on to Stalag Luft I, Barth.

The final Eagle cage was shut.

11

Evasions in Three Countries

THE HEIGHTENED AIR ACTION of 1944 also gave impetus to three adventures of downed Eagles and the heroic civilians who aided and supported them. In Belgium, France, and Holland, the three Eagles encountered the special courage of Underground resistance fighters, pursued the path to freedom, and witnessed the liberation of western Europe.

On January 24, 1944, Robert Lee Priser had been flying combat missions in Europe for a year and a half, first in the Spitfires of Britain's 71 Eagle Squadron, then in P-47s with the U.S. Eighth Air Force for seven months, and most recently, the P-51B Mustang in the Ninth Air Force. Less than four weeks before, soon after getting married, the 22-year-old Miamisburg, Ohio, native had been promoted to commanding officer of the 353rd Fighter Squadron, 354th Fighter Group.

As he flew high above German-occupied Belgium, four aircraft approached him that seemed to be friendly P-47s. Then, suddenly, there was fire. Priser described his final dramatic mission and its denouement:

> On January 24, 1944, Eighth Air Force sent its bombers to Frankfurt. We had just taken over the escort duties from the P-47s, who had turned for home, when a recall order was received. I led 353 Squadron around to follow the last of the bomber stream out when I observed a flight of four P-47s several miles inland. These I ignored momentarily, while checking visually that nothing was attacking our bombers below.
>
> As I looked back at the P-47s, the wingman to the formation leader opened fire from about a mile away and with 90-degree deflection. At this closer range I could see that the leader was definitely a P-47 and the three other planes were FW 190s with a paint scheme similar to P-47s.
>
> At this instant my Mustang shook from 20 mm cannon hits. Quickly,

I released my wing tanks and turned hard right and then hard left and got behind the enemy formation. I fired a long burst at the P-47 and then another at the FW 190 that had fired at me. I could feel my engines cutting out, so I pulled out of the shallow dive and leveled off to check for damage.

Most obvious was a large hole in the right wing root. Flames were eating everything inside. Farther out the right wing, I could see where 20 mm cannon shells had hit without exploding, and gasoline was pouring out everywhere.

As I set a course for home the fire intensified and the engine got rougher. After several more minutes I knew I would have to bail out immediately.

I jettisoned the cockpit canopy, disconnected the lines and seat belt, and jumped. Then I felt a tremendous blow in my midsection. I was draped over the left-hand elevator. Uncontrolled, the Mustang had gone into a high-speed dive. The air pressure was so great that I wasn't able to lift my arms to push myself off. Knowing that I had no chance to get myself off the tail, I said, "Oh God, help me!"

Apparently the prayer was answered because I woke up while drifting down in my parachute close to the ground. Nearby was an unpaved country lane where a man and two children, all with bicycles, watched me.

I landed easily on the soft ground, got out of my parachute harness, and took off my flying boots. The man was gesturing and pointed to a farmhouse nearby. It was obvious he wanted me to go there. I had sense enough to lay a false trail away from the farmhouse by using the coveralls, leather helmet, and boots, then doubled back to the house, which was about 150 yards distant.

An old woman met me and let me into the kitchen. I took off my leather flight jacket as she poured me a basin of water. She held up a mirror, and I saw that my face was bloody and the whites of my eyes had become bloodred. As I washed the blood from my face and hands, the man from the road walked in.

After a short discussion in French, the woman produced a civilian jacket, a scarf, and a beret. I put these on, and the man let me outside and showed me a bicycle on which I was to follow him. We rode along the dirt road for about a half a mile and topped a small hill. As we started down the far side, three German military vehicles filled with soldiers passed us from the opposite direction. We ignored them and they went by without stopping.

A bit farther along, the dirt road ended at a cobblestone lane. We stopped at a farmhouse, and my escort took me to a special hiding place to rest. I fell asleep immediately.

Several hours later the man took me into the house and gave me some food. I handed him my military wristwatch, as we had been advised to do in escape classes. The man laughed and bared his left arm. At least a dozen military watches were strapped onto it.

On the bicycles again, we followed a back road. I was feeling uncoordinated, and hurt all over. Finally, we stopped at another farmhouse and went inside. I sat in a chair, gratefully, and dozed while my guide talked with the others. Soon there were polite goodbyes, and we rode away again. We stopped at several more farmhouses for short visits. Late in the afternoon we arrived at one where I was put to bed. It must have been obvious that I was in shock.

I awoke about noon the next day when my hosts brought in some food. I held my body rigid because the slightest movement caused excruciating pain in my lower rib cage. I was sure then that I had fractured some ribs when I hit the tail of my Mustang.

My hosts were two short, chunky men, about age sixty, with large mustaches. They appeared to be brothers, and spoke no English. I never saw any women around. They must have been moved elsewhere while I was staying there.

On the seventh day I was able to get dressed and joined the two men for supper of vegetables and roast chicken, rare on one side, burnt on the other. It tasted good. From the men I learned that I was in Belgium a few miles north of the French border.

Several days later a man I came to know as Jules Rousseau arrived in a horsecart to drive me to his place. He brought along a hat that was too large for me and a topcoat. My hosts bussed me on both cheeks in the French manner, and I thanked them for their hospitality.

Driving along a cobblestone road into a town, we were stopped by a German sentry at the main intersection. After a short discussion with my driver, we were allowed to proceed. We followed a bigger highway out of town, where camouflaged German trucks passed us. A bit later we came up behind a company of German infantry. We followed their line of march until they turned off and went into a large church, which, I was told, was their barracks. Beyond that point we turned onto a side road and arrived at our destination, 35 Rue Belle Tête, Ecaussinnes. This was the only address I had memorized. I realized later that my knowing addresses could only increase the already considerable risk these people were taking by sheltering me.

During my stay with Jules, I was provided with an identity card and a work permit that stated I was a mining technician in a coal mine near Charleroi. My new name was Robert Wosswinkel. Later Jules gave me a .25-caliber automatic pistol and a pair of "brass knuckles" made out of aluminum.

After several weeks I was moved to another refuge in Chapelle. During the next four months I stayed in or near towns such as La Louvière, Gouy, Le Piéton, Rhode-St. Gènese, Braine-le-Comte, Soignies, and others. Sometimes I would stay only one night at a house, more often three or four nights, and occasionally two or three weeks. Always I was cautioned to stay away from windows and not to go outside except under certain conditions. To pass the time I often listened to the radio or read books that usually had been left behind by the British after World War I.

The food was good but limited in quantity. Most of the people I stayed with baked their own bread, rather than eat the official ration bread permitted by the Germans, a dark brown bread that tasted like it was made from sawdust and was usually fed to the household dogs.

The last week in May 1944, Maurice De Vroom took me by train to Brussels, where I spent the night in the townhouse of the physician to the King of Belgium. The next day a servant named Robert escorted me on a long tram ride to the physician's country home at Petite-Espinette, just south of Brussels. Several German soldiers were on the tram and got off at our stop.

The country home was large and sat on about two acres of land. It had a tremendous view overlooking a large valley. It was the second house from the main road. The first house, I soon learned, was headquarters for a German general and his staff. Other houses in the neighborhood were used as barracks.

A block south on the main road was a sidewalk café frequented by both the civilians and the Germans. Robert and I often sat at a table enjoying coffee, sometimes watching squadrons of USAAF B-26 bombers flying overhead in formation, unmolested except for occasional flak. German soldiers walked past our house going to or from duty next door. Sometimes at night we could hear the sound of an occasional gunshot. Often I watched German work parties next door stringing telephone wires or performing other duties. More than once a German soldier stopped me on the street to ask for a match.

One morning in July, Robert had gone to Brussels for the day, leaving me alone. I was in the basement kitchen peeling potatoes when I happened to look up at one of the high windows and saw a German's head go past. Quickly I placed the potatoes out of sight and slipped up the stairs past the first floor, and peered out the windows on the second floor. I saw six or seven Germans moving around, trying to arouse attention. They pounded on the door for a while and looked into the windows, and then gave up and returned to the house next door.

When Robert returned home that evening, I told him about the visit. The next morning he went next door to check, and then came back and told me the Germans were going to requisition the house and I would have to move immediately. That afternoon Maurice De Vroom came by and led me on foot to another place several miles away, a small summer cottage near a large forest. Here I had complete freedom to go outdoors and into the forest. It was wonderful. The people, Albert and Madelaine Abrassard, treated me like a son. I stayed with them for more than a month.

At this location a young man, Marcel Montoisey, sometimes came over to walk with me around the neighborhood, showing me the roads and pathways. He told me he had been taken to Germany and forced to work. He had escaped, but had been beaten severely. There were scars all over his head, and some of his teeth were missing. He showed me a cave he had dug into the side of a hill in the forest. He would hide there when the Germans came searching.

U.S. Eighth Air Force bombers would come over at times in massive formations, usually harassed only by flak. Sometimes at night, lying in bed, I could hear cannon fire up above, and I would rush outdoors in time to see a little glow grow larger and larger and become a British bomber falling vertically in flames. Whenever a plane came down nearby, I would hide out in Marcel's cave until the German search died out.

One day Albert and Marcel took me into the woods to a small clearing. A 1930 Chevrolet sedan drove up and five men got out. Three of them had pistols and rifles. The other two obviously were prisoners. Albert said the two had parachuted, but they possibly could be German agents trying to infiltrate the Belgian Underground. The two had dark complexions and spoke English poorly.

After some questioning, I learned that they had been gunners on a B-24 and had been shot down on their first mission. Their parents had immigrated to Hawaii from India just before the war. They had been drafted, trained as air gunners, and put on an all-India-crewed B-24. I told Albert what they had said and also that I believed them. It sounded typical of the military, and even the Germans wouldn't use a tale that stupid.

Albert agreed and told the Belgians what I had said. They laughed and drove away, taking the two Indians with them. Albert told me later that there had been a pick and some shovels in the car. Had I given the word that the men were fakes, they would have been killed and buried in the forest.

By the beginning of August, food was getting scarcer, and we lived more and more on the garden. We had a lot of strawberries, but these became tiresome. One morning Albert went out on his motorcycle and returned with a real prize—a sheep's head. I still wasn't hungry enough to eat brains, which Albert and Madelaine relished, but the rest of the sheep's head made a wonderful stew.

In the middle of August, I was moved to a nearby farmhouse, where the food was somewhat more plentiful, as most farmers had a supply of pork and potatoes. The news on the radio was better, with Patton and Montgomery advancing. One morning a squadron of P-47 fighter bombers strafed and bombed an area about a mile south of us. At noon I watched two German soldiers drive a farm cart, pulled by a balky horse, on the road past our orchard, so I knew the British were close by.

On the afternoon of September 3, 1944, Marcel Montoisey came by and said the British spearhead was on the main highway a mile away. He offered to take me there, but first had to pick up a Belgian man who had served in the German Waffen SS in Russia. Marcel had no trouble taking the Belgian collaborator except for his mother and grandmother, who became hysterical as he was marched out the front door.

Marcel, the collaborator, and I found that the British "spearhead" consisted of a lone military policeman. The MP said he would take me with him when he went forward to join his unit. Marcel said goodbye and left with his prisoner. Finally, half an hour before dark, the MP got on his motorcycle, I climbed up behind him, and we were on our way. We got into Brussels a bit after dark without seeing anything of the British army and stopped at a police station. The building was crowded with civilians wearing arm bands and carrying German rifles and grenades. Nobody knew where the British were. They mobbed the MP, as he was the first British soldier they had seen in four years.

We rode farther into Brussels and stopped at what proved to be a bar, Belgian style. It was full of patrons who also mobbed the MP, since he represented the British Army. They bought him many drinks, which he gladly accepted. Finally, I convinced him that our duty lay in finding the British. When we started off again on the motorcycle, around midnight, my MP was nearly drunk.

In the dark we got onto a main avenue and caught up with a convoy of trucks, tanks, and some motorcycles. After a while I realized that we had joined a German column evacuating Brussels. I informed the MP of this, and we turned off on a side street and headed back the way we had come. Well after midnight we encountered a British roadblock that turned out to be the spearhead we were looking for, made up by the Irish Guards division. A British officer fixed me up with a place to sleep.

The next morning, while the Irish Guards were moving into a large park in Brussels for several days of rest, three British officers took me along on a jeep tour of liberated Brussels. People lined the streets everywhere, cheering, and often mobbed the jeep. Occasionally we saw piles of furniture burning in the streets. We were informed that this was informal retaliation against persons known to have collaborated with the Germans.

Allied airmen who had been shot down began arriving at the park that now was our temporary base. One sick-looking group of about twenty men had been held by the Gestapo and were in poor condition—thin, weak, sallow-complexioned.

For some reason a British light plane flew directly over the park. Several rounds of 88 mm flak burst just above the ground, and shrapnel whizzed in all directions. Both the military and the civilians fell prone or rolled under vehicles. In spite of the crowd of civilians, few persons were wounded, none seriously.

The following morning all Allied airmen were put on trucks headed back to Normandy. We saw foxholes, dead horses, and shot-up tanks and vehicles, the residue of many small intense battles at strong points. We drove through high mounds of bricks, all that remained of one town in Normandy.

Flown back to England, I was debriefed and given a new uniform and lots of back pay. I sent telegrams to my wife in Ipswich and my family in the United States, then went on 30 days' leave. My wife and I decided to visit Scotland. On the way we stopped off in Saffron Walden, and I went to Debden to visit with the 4th Fighter Group. By long coincidence, the 354th Fighter Group was temporarily based at Debden also.

I stood in the doorway of the Officers' lounge at Debden and looked around the large room. I had never seen it so crowded. There were many new faces and some familiar ones, too. I felt as though I were an outsider.

Then a voice shouted, "You're Priser—but you can't be! You're dead! I saw your plane go down on fire and you didn't get out!"

I turned and saw that the speaker was a pilot from my 353rd Squadron. "I was draped over the stabilizer," I told him. "If you had been closer you would have seen me."

I was free after seven months and ten days of hiding, of having others risk their lives to help me. But there had been a price. I have since learned that at least four of the people who helped me were executed by the Germans.

★ ★ ★ ★

On March 5, 1944, Steve Pisanos was flying his P-51 Mustang at 2,000 feet over German-held Le Havre when the engine went dead. Pisanos could not be sure, but suspected the trouble was with the spark plugs. "We were using Spitfire plugs on an American-built engine, and they were good for only five, six, or seven hours." In any case, Steve Pisanos knew he was about to crash.

> I wanted to glide as far as I could, and knowing that the Germans would shoot at anyone coming down in a parachute I waited to get as low as possible. I released the canopy, took my helmet and gloves off, and tried to get out from the left-hand side. I put my feet on the outside of the wing, holding my dinghy cord, which was still in the cockpit behind the seat. I tried to pull the cord, but it was jammed. I reached for my regular knife but did not have it, so from my pocket I pulled a knife blade an inch and a half long that I had found on a street in New York. Thank God, I was able to cut the nylon cord.
>
> But then I saw that the plane was heading close to the roof of the only barn in that part of the country. It was about to crash on the rooftop when I jumped into the cockpit and pulled the stick enough to avoid the roof. The aircraft stalled, the right wing hit the ground, and I slid along the ground with the plane. I remember going through the prop blades, which luckily were no longer turning.

At first, the area seemed quiet, but Pisanos soon found himself under attack. His narrative of evasion continues:

> My plan was to set the aircraft on fire and get away. After I untangled my chute and eased my left shoulder so that I could move, I removed my escape kit, planning to dip my scarf in the wing tank, light it with a match, and set the fire.
>
> Before I was ready to strike the first match, I could hear machine gun firing. I was scared—down the valley I could see two German soldiers running toward me firing—and the match went out. I thought I still had a chance to set the fire with a second match, but the bullets were too close to my head then. I dropped the matches and the escape kit and started running.
>
> I ran toward the forest. I could see bullets hit to left and right of me, and I dived into thorny rosebushes bordering the forest. I was scratched all over. I had no gun, nothing to fight with. At our escape lectures they would tell us that the best way to escape after a crash was to get into a forest—but make sure you have a compass. I took the compass out of my boot, but I could not rig it. As it turned out, I did not really need it.
>
> I could hear the Germans forcing their way through the thorns and bushes, following me into the woods. I ran and then sort of circled back in a triangular or rectangular course and saw the open hole where they had pushed into the bushes. I could hear them, a way behind me, so I crawled out of the bushes, ran past my plane and past a farmhouse, and came to barbed wire along a paved road. A motorcycle was parked there. I did not know how to ride it, so I left it alone.
>
> I spent four or five days wandering around all over Normandy. The first night I slept under a bridge, and the second night in an abandoned house. Our escape lecturer had told us, "If you see a haystack, use it." I had sighted a haystack near the intersection of a Y-shaped road, so when darkness of the third night came I dug out a sleeping place in it. About 1

or 2 A.M. a German convoy came along and stopped to decide which arm of the Y they would follow. The French had taken all the road signs and turned them around, to confuse the invaders. Several soldiers took advantage of the stop to relieve themselves in the hay. I held my breath, but they did not spot my hiding place.

On the fifth day I saw a small town in the distance. Later I learned that I was in the Department of Calvados, more than 100 miles west of Paris. I stood on a little hill behind a bush and saw that there appeared to be no German traffic around. I walked down into the village. I had thought of surrendering, but then I had told myself, "Don't do that. That's not the way to play the game."

I went through the center of the village and saw no sign of German activity whatever. Very soon I noticed some people looking at me through a window. A young fellow came around a corner and asked, "Are you English?"

"American pilot," I replied.

He took my hand—I must have looked like I was very far gone physically then—and hurried me through the square to his home. His mother grabbed me and welcomed me, and right away fed me. She could see that I was half starved. I was still in my muddy jacket and flying uniform.

I stayed with this family for 15 or 20 days. They brought in a French doctor to look at my shoulder, and he said I had to have it X-rayed. The nearest city with the equipment for this was Rouen, too far away to consider, he said, but added that "I think there is a place close by that will be able to take care of you."

I could speak French a little. The next morning two gendarmes, who were really Maquis, came in a Citroen, and I rode with them and the doctor to the next village. There was a machine gun in the back of the car, covered with burlap. We stopped in front of a house with a German flag flying outside of it. The doctor went inside.

They had already prepared papers for me. These gave me the identity of one Jean-Claude Boyer, a mechanic from Marseille. They didn't want the Germans to question me, so they said I had lost my hearing and my ability to speak because of injuries from artillery fire. They said to the Germans that my aunt had two children in Germany, working for the German cause, and she had asked me to help run her farm. They added that I had climbed trees to cut branches, and fell from one tree, smashing my shoulder.

I started to become suspicious of my French escort, afraid that this might all be a put-up job. So I sat on a chair saying nothing, and my heart was going out and coming back again. A German soldier came in, took me into the X-ray room, put me on a table, and X-rayed my shoulder. Then my gendarmes bustled me out, before there could be any discussions, put me in the car, and we left immediately. The doctor remained, to get the results. He had used this setup before.

Later in the day the doctor came back to the house where we were staying. Everything was fine. My shoulder would heal in time.

It developed that the man of this house where I was staying was an intelligence officer for the Free French, the Maqui—the Underground—and he needed assurances about me as much as I needed assurance about the people who had taken charge of my life.

An unknown number of young American men—particularly men with German backgrounds or German family ties—had joined the Gestapo and were being used to impersonate American pilots and to infiltrate the Maquis, I was told. "So we need to find out who you really are," one of my French friends explained. "We have to know where you were born, where you went to school; the identity, rank, serial number of every man in your squadron. We need the name of your roommate, the letters on the aircraft you were flying, your parents' names, the name of your commander—everything that is not a military secret."

Recognizing that disclosure of this information was vital to his continued protection, Pisanos proceeded to describe some of the crucial events of his background. Born November 10, 1920, at Athens, Greece, into the large family of a subway motorman, he had left his home country at the age of 18, working as a cabin boy aboard a ship crossing the Atlantic. Once ashore in New York, he had managed to take private pilot lessons and to become certified. When the war had broken out in Europe, he had aggressively sought out an opportunity to fight and fly for the Allies, but at first had been frustrated by his lack of American citizenship. Finally, through the Clayton Knight Committee, he had been accepted for training, had qualified for operational flying duties, and had been posted to the RAF's 268 Squadron.

During trips to London, Pisanos had met Yank pilots at the American Eagle Club—including Chesley Peterson, leader of 71 Squadron. He had visited the Eagles at their home base of Debden, and at the urging of Eagle Vic France, had reported for duty with them. Several months later, when the three Eagle squadrons were transferred from the RAF to the U.S. 4th Fighter Group, Steve Pisanos had, through the efforts of Peterson and the American ambassador, John Winant, become the first beneficiary of a new federal law permitting aliens serving in the U.S. Armed Forces to become citizens, regardless of the number of years of residence. For the past year and a half, he had proudly served his adopted country as a pilot in the U.S. Eighth Air Force. He had been credited with ten enemy aircraft destroyed.

The Frenchmen seemed pleased with these and other details Pisanos provided them, and they excused themselves to check the information through their intelligence network.

The next day they walked into my room and said, "Hello, you; you are the famous Flying Greek. We talked to London last night, and we learned for sure that you are our man." They hugged me and kissed me and opened a bottle of Bonaparte wine. Now I knew that I was fully accepted as a trusted ally.

About a week later, a woman from Paris came out to look me over and to arrange for me to be moved to the city. A truck loaded with firewood, supposedly fuel to be sold in Paris, picked me up. The drivers had a revolver in the glove compartment and machine guns in the trunk. We were stopped several times by German guards, but were passed without question when we produced our papers. As we approached an air base, a German soldier came up, pulled his gun, and said he wanted a ride to Paris. Our driver pushed the machine gun into position, and the man put his gun back in the holster.

In Paris, Pisanos stayed with six different families as the diligent and intelligent Gestapo forced several emergency getaways. Once, Pisanos and a woman were trapped in the street. German soldiers had sealed both ends of the block, stopping people and demanding to see their papers.

"I'll go across the street. You stay here," the woman said. A leather-coated German demanded of her, "Papieren!" The woman responded to him slowly, as if she were hard of hearing, effectively confusing the Gestapo men. As they impatiently tried to deal with her, they absently glanced at and stamped Pisanos's *Papieren* and passed him through.

On another occasion, Pisanos and some Resistance people were moving from one house to another when they found themselves trapped in an alleyway. The police were apparently looking for someone else there, but the Maquis could not risk being ensnared in the Gestapo's net. To avoid going through the control point, the group entered a manhole into the Paris sewer system. For half an hour, they navigated through a wet and blackened maze beneath the City of Lights. In the large tunnels, they had to avoid rats 1½ feet long. Finally, they emerged in a square beyond the control area.

Yet another time, Pisanos and a fugitive RAF pilot were hidden on the fifth floor of an apartment house with a French couple. The husband was an engineer engaged in making starter motors for the Me 109, which he was systematically, though discreetly, sabotaging. The rule of the house was to go to the balcony whenever something risky began to develop. At two o'clock one morning, the police banged on the door—perhaps because the RAF pilot had been playing the piano late in the evening. The two pilots were sleeping with their clothes on and quickly got onto the balcony. To conceal their presence, they had to jump to five other balconies. At last, the French hosts let the banging police in and took them onto the balcony, telling them that "the two people you wanted to catch jumped down and took off in their cars." With the police now looking elsewhere, Pisanos and his colleague returned to the apartment.

As Pisanos's stay in France lengthened, he was drawn into a deeper involvement with the Resistance, starting as a casual spy. For a while, he stayed southeast of Paris with a doctor, his wife, and baby.

> In the evenings, I would walk along the Seine with a little local girl. She thought I was a Polish refugee. I noticed a river barge with German soldiers on the decks, along with many pipes and valves. "What's that?" I asked the girl.
>
> "The Germans are very smart," she answered. "The Americans machine-gun the fuel at the air bases during the day, so the Germans can't keep fuel there. They keep it on the barges. At night, trucks come in, load

up with fuel, and take it to the air bases. Next day, the Germans are ready to meet the Americans. Every twenty kilometers or so, they have a barge."

We went back to the house. This was a gathering place for intelligence officers; the wife was an agent. The first Americans from the OSS (Office of Strategic Services—the proto-CIA) were coming in once a week. I told one of them about the barges, and he asked the Frenchman in charge, who said, "I don't think those are fishing boats."

I persuaded the American to come with me down to the river late at night. He looked at a barge through the bushes. "That's a storage area for fuel," he said. "I'll take care of it."

Not long after, I was walking with the little girl farther down the river. On the far side I saw and heard an Me 109 engine being tested on a stand. It was new. A man back at the house said that the only factory located there made radiators for home heating. Again I took the OSS agent to the river, where he saw the engine. Now it had a three-bladed prop and a cannon in the hub.

He passed the information to London. About two weeks later, B-17s hit the barges. Reconnaissance several miles down the river had revealed an entire Me 109 assembly and testing area. The aircraft and engine parts were being brought to the river, where the engines were also tested. The finished fighters were then sent to nearby bases. That target, too, was smashed by U.S. bombers.

The Flying Greek's courage, combative spirit and cunning were increasingly appreciated by the Maquis, and it was inevitable that they would be put to use. Pisanos was drafted as a willing saboteur and bushwacker for the cause.

As the time for the Allied invasion of France obviously was growing near, the Resistance became increasingly bold, striking more openly at the enemy as they were supplied with weapons and target assignments by the Special Operations Executive Command in Britain. They also selected many targets of their own. Pisanos became a part of their campaign to spread death, destruction, and terror among the already jittery German army of occupation.

A typical operation was the ambush of a convoy. In one such case, the column of vehicles contained a staff car in which the Maquis knew some high-ranking German officers would be riding. The attackers set up their ambush around the curve of a road. Pisanos, at one end, was to throw a captured German grenade onto the lead car.

When the convoy arrived, Pisanos threw his grenade, but the pin broke, probably sabotaged by a slave laborer at the factory. Nevertheless, the attack went forward, swiftly and savagely. With tommy guns and grenades, trucks and cars down the convoy line were wrecked and their guards killed.

German transport was constantly punished. Convoys were stopped, for instance, by knives embedded into roads. The deserted cars would then be dumped into the Seine.

In July and August 1944, the Germans' worst fears were realized. Having established their beachhead in France in early June and having weathered stern German resistance through July, the Allies broke through and around the German forces and advanced on Paris.

Adolf Hitler had given orders that if the Germans would have to relinquish Paris, the city would be wrecked and humiliated in the battle. It would burn and crumble, wracked by explosions set at places whose loss would most pain the French. What the Germans were contemplating was clear. Certainly they intended to blow, among other treasures, the bridges across the Seine—bridges that were beautiful, gracefully sculpted, beloved works of art. The Underground determined to save them, and Steve Pisanos, a Greek-born American flier, was enlisted to help save part of the heritage of France, part of the heritage of the best of European civilization.

It was delicate, hazardous work. Under the noses of sentries, the Maquis had to scramble to where the boxes of explosives were set. With great care, in darkness, they had to remove the dynamite and ruin the controlling detonation wires. To prevent the Germans from realizing what had been done, they had to leave the explosives' boxes apparently undisturbed. All was done in silence and in fear, but it was done.

As American and Free French troops closed upon Paris, the Germans began to evacuate, in spite of Hitler's demands that the city be defended to the last man. The evacuation became a panic as fleeing Germans pulled horse carts and tried to drive cars without tires when normal transport wasn't available. And the Maquis always seemed to be at the next corner, on the roofs and in the apartments of buildings, firing and lobbing grenades into the Germans fleeing through the streets below and erecting barricades to slow and trap the driven enemy. In the northeast part of the city, the Germans were forced to drive pell-mell along sidewalks in their fear and haste. Meanwhile, when they reached open country, the troops were mauled by Allied fighter bombers, as the Luftwaffe had strafed and bombed fleeing civilians on the roads of Belgium and France when the war was young and the Nazis' New Order was winning.

On August 25, Paris was liberated. Thanks to the efforts of the Underground and the perilous refusal of the German commander to carry out Hitler's orders, the city did not burn. The bridges remained intact, as did Paris's other treasures. Through brave defiance, civilization defeated terrorism in the City of Lights.

* * * *

On April 10, 1944, Major Donald Kenyon "Don" Willis was escorting bombers attacking Gutersloh airdrome in Germany and helping to evaluate a new technique. It was the first operational use of the droop-snoot P-38, a modification of the Lockheed fighter in camera coverage of targets.

By now, Willis had been in the war for a very long time, having first flown open-cockpit Bristol Bulldogs, with a machine gun strapped on one side, for Finland against the Russians in 1938 and 1939. With the Finns' collapse, he had fled by dogsled team across the northern part of Sweden into Norway, then flown with the Norwegians against the Germans until that cause had been lost. In the company of a Norwegian comrade, he had proceeded to escape from Tromso, Norway, in a stolen German Heinkel III seaplane, across the North Sea to the Shetland Islands. Then, after flying with an RAF Norwegian squadron for a while, he had transferred to 121 Eagle Squadron in June 1942.

On this, his eighty-sixth mission, he was north of Rotterdam, returning from Germany, when he reported by radio that one engine had been hit by flak and he was having trouble with fuel pressure. He tried to fly out over the sea, but instead made a dead-stick landing in the Walcheren area of Holland, some 50 miles southwest of Rotterdam. With the engines out, the Germans failed to hear the plane come in. Willis came down between a sports field, where a football game was in progress, and some dikes. A wing struck the dikes and was damaged.

As he scrambled out, Willis saw spectators and players running from the football field to the crash scene. "At least 500 people were milling about the crash and along the dikes," he said later. "I ran to the path where the people had left their bicycles and took one, grabbing with it a long red coat to throw over my flying jacket and green trousers. The coat had 10 guilders and a watch in the pocket.

> I got in among some cyclists on the path at the top of the dike and pedaled toward the German soldiers who were running from a nearby gun post. The soldiers were busy for the first few minutes trying to keep people away from the plane, though several climbed on the dikes and searched the countryside with field glasses. I pedaled beside a woman who kept watching me out of the corner of her eye, but she never spoke. When we rode into a small village, the woman turned down a side street. I parked the bicycle by a stone bench and sat there trying to think out my next move.

Having rested on the village bench briefly, Willis resumed his bicycle journey, but soon noticed that another cyclist had started following him. Willis was quite sure the man had been among the spectators at the football game who could have sighted him as the crowd ran toward the downed P-38. Willis accelerated his pace and then quickly jumped off the bicycle, ditched it, and crouched down in the roadside growth. His pursuer was not to be eluded that easily. The man found him, called out that he was a Dutchman and a friend, and told Willis to stay hidden until he returned.

Shortly, the Dutchman came back with a farmer's cap and scarf for the American, and accompanied by a small boy. "This is my son," the man said. "Our name is Kuppens. We shall try to help you."

The three started walking, with the man out in front. Kuppens told Willis that he would lead the way, and if anything looked suspicious he would cross the street, as a warning to the American and the boy. Outside a church Willis and the boy sat down on another bench to rest. A German truck came up in front of the church. In the truck were several soldiers and a couple of bloodhounds.

The Kuppens boy got up and walked onward, casually. Willis stayed on the bench, his heart in his mouth, as the soldiers jumped off the truck and lined up in two ranks, facing him. The officer in charge stood with his back to Willis, ignoring the "farmer" in cap and scarf on the bench, and called out orders to search the area. One soldier would stand on the dike with binoculars while another would criss-cross the field with his dog. Others would sweep adjoining parts of the neighborhood. After one field was cleared, the search crews would move on to the next.

Willis reported:

> I walked around the little town waiting for the Germans to get some distance away before following the route taken by one of the search parties. After they searched a barn, I crawled in thinking it was the safest place to hide at the moment. I was seen by the woman who owned the barn, and she hurried out to tell me I could not stay there. I had no trouble understanding her because I spoke Norwegian, which is somewhat similar to Dutch. She promised not to tell the Germans that I had been there if they came back. I crawled down a drainage ditch to a field of high grass and hid for most of the rest of the day.

Late in the day Kuppens returned, found Willis in his hiding place, and walked him to a small workshop behind his modest home. He hid the American under the workbench where he fashioned wooden shoes. Kuppens told Willis the Germans had been searching the whole of the village, all day, and had been questioning everyone about the pilot known to be in hiding somewhere in the vicinity. The Dutchman added that, with darkness, a total curfew would go into effect. There would be no chance to move on until daylight the next day. Meanwhile, Kuppens promised, he would make the visitor a pair of wooden shoes, and Mrs. Kuppens would gather up what remaining flour she had and bake bread that he could carry on his way.

The kindly hosts fitted Willis with Dutch civilian clothing and gathered up his American clothes and buried them—except for his boots. After second thought, Kuppens realized that the flier never would be able to walk far in clogs. It was decided that he would keep his boots, even though they possibly could betray his identity.

Next, Willis examined his Escape and Evasion Kit. For navigational needs, he carried cloth maps of France, Holland, and Spain, and a compass. He had enough Horlick's malted milk tablets and chocolate bars to sustain him, if properly rationed, for four days. He also had matches, adhesive tape, and chewing gum, and three escape photos. He carried a purse containing 2,000 French francs, to be paid out to the people who helped him.

So armed, Willis headed for the city of Brussels, 100 miles away. At one point he approached what turned out to be a pillbox and heard German conversation within. Quickly, he backtracked to a pasture and found a cow, and boldly drove her past the sentry box—to all appearances a Dutch herdsman retrieving wandering livestock.

Another time he saw a German guard at a bridge ahead. Willis took from his pocket the remaining half of a cigar he had broken in two in his crash landing and stuck it in his mouth. Whether the unlit cigar really helped give him the appearance of a genuine Hollander or merely bolstered his own courage, Willis crossed the bridge unchallenged.

Willis decided that, curfew or no, travel by day was unduly slow and risky. When conditions were favorable, he would also move by night. As he later wrote in his official Escape and Evasion report:

> After dark I checked my compass and walked cross-country in a SW direction. It was not easy walking because of the many dikes and fences. About midnight I stopped to rest at a barbed wire entanglement and was just missed by three German soldiers walking along a footpath. I thought they were a searching party, and I didn't move until I discovered they were the relief for a small gunpost that I had nearly stumbled into.
>
> I went toward a group of buildings to see if I could find a place to sleep. A young boy and girl hailed me as I started to crawl into a haystack, and after some difficulty I convinced them I was an American airman. I waited at the haystack while the boy took the girl home. He returned with food and said he would walk with me. He knew the country well and by 0800 hours we had reached Roosendaal (a town about 40 miles from Walcheren). My friend turned back then after giving me excellent advice and directions. There were some German strong points to be avoided. I was shown how to go around these and I was helped by traffic signs put up by the Germans.
>
> I found the road to Antwerp and followed it, keeping to a safe distance in the fields. Around noon I reached Esschen. (He had just crossed the border from the Netherlands into Belgium.) I had passed many people but so far had not spoken to anyone. At the edge of Esschen I saw Germans on duty at a control post and stopped an old woman to ask her how I could pass this barrier. She motioned that I should go around the control.
>
> I went into a wood for the afternoon and hid where I could watch the people working in the fields. When they finished work I followed some of them along a small road into the town. There was either no control post here, or the workers going to their homes in the evening were not checked. I found the railroad on the other side of Esschen and walked parallel to it until dark, when I made my bed in a strawstack.
>
> I walked steadily all the next day, avoiding towns and speaking to no one. Twice I ran into German officers hunting in the fields, but each time I hid before they could see me. I arrived at Antwerp in the afternoon and tried to get around it before dark, but I picked the wrong direction and ran into water. In retracing my steps I walked into a German AA battery, and the soldier on sentry duty motioned me away. At dark I crawled into a haystack but couldn't sleep because of the cold.
>
> Before daylight I went up to a farmhouse and got the farmer out of bed, but he wouldn't open the door. Finally he said to come back in the daytime. I waited for him to get up because there were no telephone wires to the house and I didn't see how he could notify anyone of my presence. He gave me some food and asked no questions.
>
> I went into Antwerp that morning, having decided to ask someone

how to get through the town. I stopped at a store where a man was loafing in the doorway, and after I started talking he asked if I was an American and took me inside. He gave me 230 francs and, after assuring me that he liked Americans, told me I could ride a streetcar to the southern end of Antwerp for one franc and without too much risk. He gave me good directions for doing this, and when I had gotten outside of the town I stopped at a roadside cafe to get some beer. There was a sign advertising Bock beer, so I put down a franc note and said "Bock." The few people in the cafe paid no attention to me, but the Belgian who gave me the beer guessed my identity. I was taken into a back room and given some eggs and bread, but not a word was spoken to me. Just as I was leaving the Belgian brushed off some straw that was clinging to the back of my coat and smiled while doing it.

I started to Brussels, following the railroad line, but got on the wrong tracks. Because I had more confidence now and felt some desperation about getting help, I went into a small railway station and tried to buy a ticket to Brussels. There was only a porter in the station, and he explained that I was following about the only railroad in Belgium that didn't go to Brussels. He wanted to know if I had parachuted, and told me he knew an English-speaking person. We went to see this person, who said I would be helped, but while I was eating someone came in and said the police had been informed. I ran out the back door and got away from the place without seeing any activity.

I arrived at Boom (a town some nine miles south of Antwerp) about mid-afternoon. The railroad and road bridges were controlled by sentries, and many people were being checked as they crossed. I was afraid because only those people recognized by the guards did not have to show papers. Not having found a way to cross by dark, I found a haystack to sleep in that night.

The next morning I watched some laborers carrying poles across the bridge. After they had done this several times, I saw the guards were not paying any attention to them. My opportunity came when two women stopped to talk to the guards, and while their attention was diverted I went up to one of the labor groups and hitched onto the pole they were carrying. The men looked at me but said nothing.

I walked to the woods where the men were stacking the poles and found that their work was directed by a German soldier with a rifle. At first I thought he was a guard, but he was paying too little attention to the workers. I wasn't sure how I could get away without arousing the German's curiosity, and was pretending to work when a peddler arrived with an ice-cream cart. I crowded around with the workers, put down a five-franc note and walked off while the German was arguing with someone and had his back to me.

In the afternoon I stopped at a small café to buy a glass of beer and was followed out by a man who waited until we were alone before asking if I was not an American. When I admitted it and asked him what he was going to do about it, he said he would like to help. From there, my journey was arranged.

On his long trek from Walcheren to Roosendaal to Antwerp to the outskirts of Brussels, Willis had found that haystacks and barns were the poorest of accommodations for chilly nights. He learned whenever possible to steal into a railway depot after dark, when the curfew had closed down the day's activities, and find the best hidden bench in the darkest part of the station under which to catch up on his sleep. Sometimes if a train were late he would find the station harboring people overnight because they had been caught in the curfew. This meant another chilly night out of doors for the major.

Willis said his closest call of all came as he waited for darkness to make it safe for him to enter a depot, and a German officer came up and asked for directions of some kind. Willis replied in French that he did not speak German, and the officer started to rephrase his question in French. "I thought for a minute that I had bought the farm," Willis told his wife later, "But a dear little old lady who had been watching us walked up, grabbed the German's arm, and gave him his directions. The German clicked his heels, and I got out of there."

Willis arrived in Brussels hungry and completely exhausted. For most of the way he had existed on about two ounces of black bread a day and what wine he could steal from farm workers. He was afraid to drink water most places along the way, for if it was contaminated and made him ill his chances of evading capture would be gone. It was early spring and the farmers were preparing their fields for crops, and their wine in goatskin pouches hung from many a convenient fence. This Belgian custom helped the American fugitive survive his ordeal.

In the Belgian capital, Willis found himself in a changed environment. As he walked down a boulevard, a woman approached and said, "I know that you are an American. I am a prostitute, but I am not going to sleep with you. I only sleep with Germans. I have VD, and I try to give it to them. That is my war effort." The woman introduced him to her friend in the Resistance movement, Monique, a nurse in a Brussels hospital, who cooperated in hiding Allied evaders and escapers.

Monique was a pseudonym used by Yvonne Bienfait, who risked her life by making her small apartment a refuge for war refugees sent along by Underground organizations. Don Willis stayed in Monique's "safe house" in Brussels for about four weeks, along with another American pilot, Lt. Col. Thomas H. Hubbard of Fort Worth, Texas, and two British fliers.

Eventually, with the guidance of the Underground, Willis walked on through Belgium and France. There was always the possibility of treachery. At one house in which he was sheltered overnight, he was told the next morning by one of the occupants, a young girl, that her father had gone to tell the Germans he was there. The Nazis offered substantial awards for such information. Willis departed quickly and was not caught.

Leaving France, he then walked across the Pyrenees into Spain. In Pamplona, from a telephone booth, he called first the American consulate, getting no answer, and then the British consulate. The latter advised him that telephones in Spain were tapped, that the military listened into conversations, and that very possibly when he left the booth he would find officers waiting to arrest him. If the Army got there first, he might be turned back to the Germans. The Air Force would be more lenient.

Recognizing that his fate was in the hands of the Spanish authorities, Willis soon turned himself over to the local police chief of San Sebastian—with good results. He spent about two weeks in jail there waiting for a plane to England. On most evenings he would dine with the police chief, who had a way of indulging in too much wine. Willis said that usually after dinner he would take the police chief back to jail, put him to bed, and then put himself in his own cell and go to bed.

Finally, Don Willis caught a plane ride to England and freedom. In his hazardous, deadly game of hide and seek, he had made his way more than 700 miles to sanctuary. On June 28, 1944, the only American fighter pilot to be shot down over Holland and evade capture telephoned his wife Patricia from London to report the happiest of news, "I'm home!"

12

The Last Ordeal

AS 1945 ARRIVED, the days of the Third Reich were clearly drawing to a close. With western Europe back in the hands of the Allies, and Russia moving forward on the eastern front, it was only a matter of time before the Eagles caged at Sagan and Barth would taste freedom again. But first, the men of Stalag Luft III would have to endure one last ordeal—a forced march in the dead of winter, followed by desperate hunger and grotesque living conditions. This was an ordeal that tested the Eagles' very will to live. Yet, they persevered these extreme hardships to emerge once more as heroes and survivors.

In January 1945, the Eagles at Sagan anticipated a possible change of scenery. It was generally believed among the POWs that the Germans intended to take them to a mountain redoubt to be used as hostages during final surrender negotiations. In any event, it was clear that the Germans could not risk the liberation by the advancing Russians of 10,000 POWs who could get back into battle in the final Allied onslaught. Accordingly, as Charles Cook recalled, the Eagles caged in Sagan were physically preparing to undertake a move within Germany.

> Early in 1945 the camp buzzed with rumors that the Russians were only 20 miles away and that the Germans were planning to move the 10,000 prisoners in the Stalag Luft III compounds at Sagan to a more stable area.
>
> Our colonel had everyone doing 25 laps around the compound, getting in shape for a long hike somewhere.
>
> One day the Germans told us to be ready to leave on three-hour notice. There were five camps of us, with 2,000 men in each.
>
> On January 27, a Saturday night, when most of us were at a camp show—*You Can't Take It With You,* of all titles—the camp commander

came on stage and said we must go to our quarters at once and get ready to move on short order.

Bill Hall specified that it was about 8 P.M. that night when Colonel C. G. Goodrich, the senior American officer at the South Compound, gave the order to pack and be ready to evacuate.

The Russian Army was only 16 miles away, and the time to march west had come. The South camp was the first to move out, around midnight. Hall and Bill Geiger, considered unable to walk the nearly 75 kilometers that lay ahead, fell out to sick bay. George Sperry was in the first group to face the brutal weather of late January:

> They ordered us out in the bitterly cold night, to march to a safer location, and they let us know that any who tried to escape would be shot.
>
> We of the South Compound led the line, with our friends in the other camps calling out encouragement as we set off in deep snow with a blizzard starting to rage. Pausing at half-hour intervals for short rests, we walked the night through. About noon the following day we halted in a small town, Grosselten, 20 miles from Sagan, and were permitted to find places to sleep in barns. I discovered a pigsty occupied by a sow, and offered to share the space with others but found no takers. I snuggled up against the sow, who was certainly warmer than I was, and slept for several hours.
>
> The Germans ordered us out on the road again that evening. As the night wore on, many of the men were unable to continue and were picked up by wagons following us. We arrived in the town of Muskau, some 16 miles from Grosselten, in early morning, and were allowed to sleep on factory floors or in other shelters. About one-third of our 2,000-member POW group was housed in a glass factory blessed with hot water so that we could bathe and shave and do our laundry. I was surprised to find myself still in fair shape except for swollen ankles.
>
> All during the following day it was a matter of cooking any food left in our packs, comparing notes on the march so far, and checking the men from our barracks to see whether they were in condition to march again if the Germans insisted on continuing the evacuation towards the west. I remained in good condition.

Don "Snuffy" Smith's flight to Muskau from Sagan was equally harsh:

> About 10 o'clock on the night of January 27, 1945, the Germans notified us that we would be moving out of Stalag Luft III on foot at five the next morning. We had a clandestine radio in the compound, and at midnight each night we received a coded special POW message on BBC radio. We learned that the Russian Army was only 16 miles away from Sagan, and we were instructed to cooperate with the Germans and move to the west to avoid falling into the hands of the Russians, if possible. We believed that our government did not trust the Russians, and also were quite sure there must have been contact with the German government about some form of German surrender and about care of the thousands of POWs.

The Germans said the building housing the Red Cross parcels that made up our food supply would be opened to us, and we could take anything that we thought we could carry. There was no time to dispense the food to the starving German civilians. What Red Cross parcels that were left over would be burned, to keep the food out of the hands of the Russian military.

After being up all night getting our gear together, we marched out of the prison Sunday, January 28, 1945—about 2,500 POWs in below-zero weather. We reached Halbau, about 10 miles away, at 3 P.M. and stood in the town square for about two hours while the Germans were finding a place for us to spend the night. We traded cigarettes for sleds and hot water for brewing bouillon cubes, and we just about froze to death.

Finally they herded us all into a church, more than 2,000 of us into space meant for no more than 500 people. At least body heat kept us warm. The next morning I heard one POW say, "This is one time that I stayed awake in church."

The next morning at nine we were on the march again, many of us pulling our supplies on sleds, about 10 more miles to Stellesinger. Here we were put in barns and stayed two nights and one day, cooking one good meal by taking over a woman's kitchen and greasing the old man with a few cigars. British POWs from Sagan marched past us and kept going west, cold as hell.

On the last day of January, we marched more than 17 miles to Muskau, which is 80 miles southeast of Berlin. We were just about finished when we got there. We were housed in a pottery factory, a warm place, and issued bread, the first food from the Germans since we left Sagan. We stayed two days and rested, and the POWs built a little paradise for cooking.

For Charles Cook, the forced march in snowstorms and near-zero temperatures exacerbated an infection he had contracted while washing dishes at the camp. In fact, he had cut his left hand on a tin-can lid and developed blood poisoning. Now he lay in the hospital at Muskau—his arm, and even his very life, threatened.

My arms were dead and my hands frozen. Someone took me to a hospital at Muskau on the German-Czech border. They had no room. Sperry and Stanhope helped me across the street to another hospital.

My left arm was swollen as big as a balloon. A doctor cut slits in it to drain the pus. The next morning six of us were placed in a truck to go back to Sagan to the Stalag Luft III hospital.

We were all but dead, six of us. A padre gave us the last rites. Then a surgeon from South Africa looked at us. He had a small vial of penicillin pills, and he distilled water and crushed the pills and gave us two cc's every half an hour until it was gone.

They were ready to bury us, but that medicine pulled us through. I was as close to death as a man can get without dying. I seemed to be floating on a white cloud. There were angels around. Nothing could hurt you.

When the penicillin took hold, my arm really got sore. Now I can't take penicillin any more. I have to take streptomycin instead.

A fellow prisoner of war, shocked at Cook's pitiful condition, took the trouble to make sure that the pilot's family at least would have some information about him. He found a scrap of paper and an envelope, printed in pencil this succinct note, and saw that it was mailed to Cook's address in California:

> Infected arm. Blood poisoning at Muskau Gr Hospital Jan. 30th on the march. Had chance of living. No info available afterwards. Russians were near.

The message bore the signature of 1st Lt. Vernon D. Adams, 1704 West Virginia Avenue, Bethesda, Maryland.

For George Sperry and others able to keep going, the next few days were a living hell as they were transferred hundreds of miles through Germany.

> We spent another night in Muskau, leaving early the next morning. It had stopped snowing, but the wind was still blowing colder than all hell. The roads were jammed with refugees—mostly German civilians fearing conquest by the Russians. After 16 wearying miles, we spent the night in barns at another village, Graustein, and next day reached our destination, the rail line at Spremberg. Here we were given our first hot food of the journey, barley soup, and then crowded into freight cars, 50 men to a car.
>
> The next two days and nights were sheer torture—locked in frigid cars without room to lie down or even sit, thirsty and starving, with only one stop, at Regensburg for water, when we were allowed to get out for the first time in 24 hours to relieve ourselves. It was simply degrading here, as men squatted in the snow with their coattails pulled up over their heads. We hurried back into the cars and the train moved up again, arriving in Moosburg that same evening.
>
> Switched to a siding, we spent another cold night in the train pleading with the guards for water, begging for the removal of sick men to a hospital, seeking any excuse to get out of the filthy, smelly cars for relief. The noise and the odors of the sick and the retching prevented even the weariest among us from sleeping.
>
> Next morning the boxcar doors were opened and we marched immediately to Camp Stalag VIIA at Moosburg, Bavaria, about 40 kilometers north of Munich. We were a sad lot of kriegies.

The train ride was doubly treacherous because the Allied bombers were pounding major German population centers. Snuffy Smith remembered the narrow escape from Leipzig:

> The second night on the train we were standing in the railway yards in the center of Leipzig at midnight when the sirens sounded and the train pulled out of town. The next morning we heard that Leipzig had been almost wiped out by British bombers, and the streets were running in blood, with some 300,000 people killed.

Upon arrival in Moosburg, some of the exhausted POWs, such as M.E. Jackson, who had contracted pneumonia, were placed in a hospital. For those who had enough strength, they soon found that Stalag VIIA was the most miserable camp of all. George Sperry described the horror of the conditions the Eagles encountered at Moosburg:

> The barracks had been hastily vacated by British officers and Italian NCOs in order to make room for us. The buildings were dirty, populated by millions of fleas, bedbugs, and lice. This unbelievable number of vermin manifested itself mostly at night. Bites covered every part of the body, including nose and eyelids. These conditions were alleviated somewhat when the Germans finally allowed us to take showers and use delousing powders, but it was not until about the end of March that shipments of American insecticide powder came through from Switzerland.
>
> The greatest problem facing us kriegies was the shortage of food. Shipments of Red Cross parcels were few and far between. Outside of a reduced ration of German black bread and potatoes, the daily soup ration consisted of a sickening concoction of ill-tasting, worm-ridden dehydrated vegetables, which we had to call green death, and grey death, a flat insipid flour soup consisting mainly of water. This time remained high on my list of incidents to be forgotten.
>
> Within a couple of days of our arrival at Moosburg, misery, diarrhea, and dysentery ran rampant through the camp. A long slit trench in the open area of our compound served as our only latrine. This area, lit up by lights from the guard towers at night, was in constant use. Day or night, rain or shine, there always was a line of men waiting for their turn.
>
> Another problem arose when the Germans stopped giving us fuel. Parties were organized to go under the buildings and rip out the underlining of the floors. When this supply of firewood ran out, wood scavenging came into its own. Barracks, latrines, and washhouses were the first to suffer. Wood also was brought into the camp by Russian prisoners and traded for cigarettes and food.
>
> For many of our men it was just like starting all over again as new kriegies. Tin cans—Klim cans, especially (Klim was a form of dehydrated milk)—were made into pots and pans to heat and cook our food. The kriegies' most popular occupation became the chow combinations and cookings of combines of two or four men. These groups could be seen in nearly every open space.
>
> By the end of February, the camp ran out of Red Cross parcels. The increased bombing of German railroads by the Allies had cut off supplies from Switzerland. Late in the month, arrangements were made to allow POWs to drive large, white-painted trucks from the Swiss border to bring in new supplies.
>
> Meanwhile the green-death soup made its appearance again. Typical of German efficiency was the way they deloused us periodically, then put us back in dirty barracks to sleep on old mattresses and lice-infected straw. Powders passed around and sprinkled on bed clothing didn't help too much. Men hung their blankets out each morning and would not take them in until night. Even then when we stripped for bed our bodies were

livid with welts. I solved the problem for myself when the weather turned warmer. I slept on the ground outside.

During the month of March we had many all-day appells, and while we were at these parade counts the Germans searched our compounds for many missing men. During these day-long periods we often would lie on our backs on the parade ground and watch formations of B-24s and B-17s from Italy flying overhead as if they were on practice missions back home. As added attractions sometimes P-51s and P-47s would come down and shoot up some of the neighboring targets and before leaving would pass over and waggle their wings at us.

Barry Mahon and Leroy Skinner, who had teamed before in escape attempts from Sagan, responded in a similar way to the deplorable conditions at Stalag VIIA—not so much to return to England, now that the war was nearing an end, but just to get away from the misery. Mahon recounted their adventure:

We were placed in a compound within a compound, with enlisted men billeted in the outer compound and officers in the inner section. Our plan was to get into the enlisted men's compound and then ask to be taken on a work detail, and to try and escape from there. Skinner and I figured that the best way again was out the front gate. One night while the guard room was unoccupied except for one guard, we went in with a pretext of making a report. The guard kept his keys on the table in front of him, one key being to the gate leading out of our compound to the enlisted men's compound.

We pushed a bar of soap down on this key while the guard was looking at the report, and made a crude impression in the soap. Later we made an opposite mold in another piece of soap, and put the two together as a complete mold. Then we poured tinfoil that we had melted into the opening. When it cooled, we had a key to the front gate.

The next night, having had a friend divert the guard, we unlocked the gate, slipped out into the other compound, locked the door, and threw the key back to a friend inside. We were still prisoners of war, but now we were enlisted personnel, rather than officers, and the Germans did not watch them as closely as they had watched us.

We approached the sergeant in charge and asked for the work detail next morning in Munich. We left at 6 A.M. on the train, divided into groups of five or six men with one German guard. Skinner was in a different group than I was. We were assigned to cleaning up bomb debris around an apartment building. I had brought several cartons of cigarettes, with the packages sewn in the lining of my French overcoat, for which I had traded my GI coat in order to look more French.

When France fell, the French army had been brought back into Germany and was used for agriculture or for work around the town. They were allowed freedom to come and go as they wished, so long as they stayed within a radius of 30 miles. So once we could break away from the guard, if we looked like we were dressed in the French manner, nobody would bother us while we walked around the countryside or in the town.

I got away by using psychology on my guard. I asked if he would like some cigarettes, and he said yes. "But it would be against the law for me to give them to you here in the open, while the Germans could see us," I said. "Why don't we go into the washroom of this apartment building and I can give them to you then?"

While we were in there I handed him a cigarette, and a package also. I felt that he would take the time at least to smoke the single cigarette, so I lighted it for him. The Germans had been confined to a ration of two or three cigarettes a day of the Balan type tobacco, and to get a good American cigarette was quite a pleasure. The other Americans and the British enlisted men working there would not give the guards anything because they hated them. So to find someone willing to share not only a cigarette but a whole package—he thought I was his long lost friend.

"I'll go back to work so they won't suspect anything, and you stay here and finish your cigarette, and that way nobody will ever know," I said. I imagine that is what he did. As I went out of the washroom I left by the front of the apartment instead of the back, and stepped out on the street as a Frenchman. I put on my beret, and that, along with the rest of my outfit—GI pants, GI shoes, and a French GI overcoat—made me French.

I walked around Munich a bit. We had been given the names of several places that would be helpful. One of them was a house of ill repute run by the French and catering to German officers. I located this address and went in the door. They accepted me, and I stayed there for a while trying to make contact. They seemed nervous about my presence. Beginning to fear that they might be in a position to turn me in, I said I had to go and buy some food, and I left, never really knowing whether I would have made my Underground connections there or not.

More or less by accident I found the railroad station on the outskirts of Munich. As I walked down the road I ran into two other officers who had escaped a couple of days earlier. One of them, Jack something, had been on this same route in an earlier escape and knew some of the places to go to try and get through to Switzerland.

We had been given quite an amount of reichsmarks from the escape campaign kit. Undoubtedly they were counterfeit—bogus money smuggled in through the British and French Underground. Nonetheless they were spendable.

The war was within a month or two of ending. Security in Germany was rather lax, or else we never would have gotten away with our trip of the next four or five days.

We would get on a train, asking for tickets in our far from perfect German. Since we were workers, we were only allowed to travel within a 30-mile radius. We would get off the train, go into a bar, order beer and some pea soup, and then we'd sit around in a ring until we could buy a ticket for the next 30 miles.

Finally we came to a farmhouse concern which had tremendous dairy and cattle-raising facilities. Our friend Jack had been there once before, knowing that there was a member of the French Underground who could get us into the connections we needed to keep going.

We got off the train and went to this place, and they gave us drinks. They told us another officer was coming in from the railway station. It turned out to be Leroy Skinner, who had gotten away in Munich the same time we did and had worked his way down to the same farmhouse concern on the Underground map.

We were happy to see him, and continued on as a group—Skinner, Jack and I, and the two other officers. We went with the French, stopping every 30 miles for dinner and to stay with several Underground people. The trains in this area were overcrowded with soldiers and evacuees, people trying to flee from the Russians or the Americans—they really didn't know which way they were going. We amused ourselves helping old ladies with their children and baggage, riding on top of cars, and really having a ball. We were able to make it to about 15 miles from the Swiss border to a place in Bavaria called Kempten, near the famous town of Oberammergau.

At this point we were told to go to a certain restaurant and order a certain dish, and when we left to say that Monsieur Lavasseur would pay. We ordered the dish and ate dinner, and when we left we said the words that were supposed to get us into connection with the Underground. Instead, they said they did not know the man, would we please pay. So we paid, figuring they would follow us and let us know we were too obvious. We walked around town until night, but nobody contacted us.

We went into another restaurant filled with French people. Finally one person came over and said, "Please don't speak English. What can we do for you?" We told him, and he took us out across a bridge guarded by German soldiers to a deserted barn and told us to stay there. He came back the next day and said he had made contact with black marketeers who would drive a truck across the border to collect cigarettes and chocolate. If we got on the truck, we would be able to cross the border undetected.

The man said he would be back that afternoon. He did not come back. We began to think that perhaps he had turned us in or had been captured himself by the Gestapo. It probably was just a matter of making the right connections, and had we stayed there everything would have been all right.

Instead, we left the next morning on our own, walking down the street toward the border. Within a mile or so of the border, we were stopped by some guards. They said that if we did not stop, their dogs would kill us. They called the dogs down from a hillside, and we could see them running through a wheat field, so we decided to stop.

Here again, the guard was amazed to find that he had captured some American officers. He said he had thought we were Germans trying to leave the country, and if he had known we were Americans he probably would have let us go. Since maybe someone already had seen him, he would have to take us in.

The guard took us back to Kempten right in the middle of an air raid. When we walked into the city, P-47s were strafing the streets, and B-25s were bombing military establishments. Apparently Kempten was a rest place for soldiers from the East front. The soldiers were there en masse, and we—the RAF and USAAF—decided to destroy as many as we could.

This little guard pushed us into the police station. In Bavaria police stations are built on the second floor with a stairway so that in snowy weather you can still get in. The guard pushed us under the stairs into the summertime entrance, and it was crowded with people trying to hide there.

An SS officer came running across the street, barely being missed by machine gun bullets. He said he had just made it by the skin of his teeth, and his other comrades had been killed. Then he began to realize that four or five Americans were standing right behind him.

He pulled his gun, turned to me, pointed it at me, and shouted "Those are your comrades up there. You are going to die for what they are doing." I thought that was what was going to happen, but strangely enough, the little guard stood in front of him and said, "Sir, you cannot shoot them. They are my prisoners. Until I deliver them I am responsible for them."

The SS officer slowly put the gun down. When he had cooled down he apologized, and said he had no right to do what he did. We took our wings off and threw them in the back of the coal bin in the police department, and from then on we told everybody that we were out of the tank corps, rather than aircraft personnel. We were put on the train back to our camp. Again this was a laugh because the guard didn't care if we jumped off or not. No one really cared.

We had heard that we were going to be liberated in a week or two, so we were anxious to get back to camp and out of the turmoil. We amused ourselves by talking to the German soldiers who were going up to meet the Americans in the last push to the Rhine. They were all youngsters. Most of them spoke some English. They were unsure of themselves, and said that when they got up to the front all they were going to do was give up and not try to fight. They wanted to know what they should do and say.

In the beginning we were straight with them and told them merely to put their hands up and say "I surrender." Soon we got bored with that, and we'd tell them to say: "Yankee son of a bitch, go to hell." We told them that means "I give up." They would copy it down in phonetics, and say it over and over. I often wonder how many surprised GIs there were who found the German soldiers running out with their hands up and saying that little slogan.

Morris Fessler, too, left Stalag Luft III in Sagan in the early hours of January 28, 1945, but his odyssey took him, Nat Maranz, Bill Nichols, and others toward Lubeck, in northern Germany. He recalled that the initial march took its toll on the Germans as well as the prisoners:

We left Stalag Luft III with perhaps one foot of snow on the ground. The Russians were only 20 to 30 miles away crossing the Oder River. The sound of battle could be dimly heard. We marched out carrying, pulling, pushing on home-made sleds what we could of our personal belongings. Temperatures were around 15 to 20 degrees. A straggly bunch we were, dressed in all sorts of clothes. Our guards were now mostly old men just pressed into service with but a few of our regular officers and a few ever present SS guards attached. At one point in the second day of marching in the snow, the old guard by me collapsed walking beside us. He was a local man of about 70 years. We lifted him up and supported him and carried his rifle to the farm where we spent that night. Next day he was no longer with us.

After three days, Fessler and his group reached Muskau, where they rested for a few days awaiting rail transport toward the Bremen area. When the transportation came, it was a nightmare like that of the Moosburg ordeal:

> We were packed in like sardines. They opened the doors three times a day for soup and water and to allow us to relieve ourselves. With the winter, the temperature, the illnesses and weaknesses of the POWs, some died en route. It was a particularly nightmarish period. When we scrambled out to relieve ourselves, of course, we had to "go'" right on the ground beside the rail cars. The Germans viewing us, both civilian and guards, called us *Schweinhund* ("pigs"). I'm sure we looked it, and I know we felt like it, too. Some, of course, couldn't contain themselves and "went" uncontrollably within the cars. What a smell and what a mess!
>
> We arrived in two or three days at Milag-Marlag Nord near Bremen, a Marine POW camp. This at least was a compound, and we quickly resumed normal POW life. As the western front came nearer, we were again on the march, this time eastward. We lived really like gypsies—mostly off the land and farms as we passed through, and it appeared our guards were living much the same way at that time. Only occasionally did the Red Cross supply us with food parcels. The internal transport system was a mess now.

Fessler's route took a few twists of its own.

> We as POWs could escape from the column for a day or two without much trouble, especially with a little help and diversion from the other POWs. There was no intent at this juncture of the war to attempt to escape back to our lines because it was too dangerous. If you were caught by the Germans, the SS would shoot you. If you managed to get past the Germans, the advancing Allies could well shoot you before recognizing you. Our goal was simply to stay alive and healthy now until liberated. At this stage we might take a day or two away from the POW column, but we would then rejoin them.
>
> On February 3, 1945, the column arrived at Tarmsted near the Elbe River. On February 6, we left there heading toward Lubeck. We arrived on the west bank of the Elbe just outside Hamburg at Blankeniste and there waited for a ferry boat to cross the Elbe the next day. We arrived about noon, and within one hour some of our POWs had discovered a rail tank car about a block away which was full of alcohol. Word spread very quickly, and in a couple of hours most of the POWs and many of the guards were roaring drunk! The German officers could only throw up their hands in despair.
>
> That evening, Nichols, Maranz, and I spent over an hour in the local pub together with some of our guards drinking Schnapps and beer. Even though we all realized this was pretty dangerous and that we could get ourselves shot, we simply couldn't resist the temptation to do it.
>
> Next morning, most had terrible hangovers from the bad booze in the tank car—which was probably a poisonous undistilled alcohol. But about 9:30 to 10:00 A.M. the first batch of POWs to ferry across started. Nichols, Maranz, and I were to ferry across in the second group. We arranged with some of the other POWs to provide us with a diversion on the other side, if the opportunity presented itself, so that we might escape for a day or two.

Most all of us were dressed like foreign laborers. After we crossed the Elbe to Blankeniste on the ferry, we all disembarked and started up the steep hill into town. As we got into the edge of the town, there was a waterfront street running off to the left paralleling the river, and as we came up to this corner the street that we were on angled off to the right and then again proceeded uphill. At this intersection the POWs nearby started a momentary diversion attracting our guard's attention. The three of us slipped a few strides away from the column and headed off down the street. There were no shouts of alarm from the guards. They had not seen us leave the column!

We proceeded northerly perhaps three-fourths of a mile to the outskirts of the town without any challenges whatsoever. There we came upon a German fellow about 20 years old working on a boat in an open garage beside the road. We stopped and started a conversation with him in German. He asked us who we were, and we told him. He said he had heard we were here and that we could be ferrying across today.

After another few minutes of conversation, he invited us to go up to his mother's home for something to eat and drink. He said to tell her he had sent us and that he would be up shortly. He directed us farther along the road, about another couple of blocks, then we went straight up the steep bank through the trees and bushes to a large home up on top of the hill.

The trio of Eagles approached the house from the protected rear and knocked on the kitchen door, unseen by any other person. A very pleasant woman in her early fifties greeted them—the young man's mother, Fessler said.

She spoke English perfectly, having been schooled in England. We introduced ourselves and told of our conversation with her son. We sat in a protected tea garden out of view of anyone. Her daughter, about 20 to 25, joined us a few minutes later. We were having tea and cakes and some wine when the son arrived about 20 minutes later. We all talked in English for about another hour and had a very pleasant time. We learned the husband had been an executive in one of the larger Hamburg steamship lines; the mother had been educated in England; the daughter worked for the Swiss Red Cross locally; and the son had somehow successfully evaded military duty.

We considered asking these people to consider putting us up and hiding us until the war was over, but we decided against it. After being there perhaps 1½ hours we thanked them, bade them adieu, and decided maybe we ought to head back to rejoin our column of POWs.

Just past noon, the three of us left the estate heading back toward Blankeniste (southward now), this time along the ridge top instead of the riverfront road, which we had come on. Several blocks down the road, which was somewhat winding, we could see ahead a Y in the road with what appeared to be a sentry box at the junction. The three of us debated whether we should evade the area. Having been somewhat fortified with wine and pleasantness for the past 1½ hours, we decided to brave it out and continue down the road. It was a bad mistake.

As we approached the Y, the sentry, a red-faced, red-haired corporal, came out of the box to meet us with his rifle in hand. He must have been watching us approach and immediately shouted: "*Gefangenen*," which means "POWs." He had us figured! Immediately his temper started rising, and his shouting became louder and louder and his face redder and redder. The three of us, Nichols, Maranz, and myself, each instinctively felt this guard was momentarily going to start shooting and that we would have to jump him and attempt to disarm him or he would kill us all.

At this very moment down one fork of the Y came an open pick-up type delivery truck and the driver, a German Army Oberfeldwebel (a sergeant) shouted to the sentry, "What's going on here?" This distracted the sentry's attention from us momentarily, and it also saved our lives, I feel. The Germans being sticklers for rank, the German truck driver took us into his custody—much to our relief.

This driver was the absolute opposite of the sentry. He told us that the sentry was a nasty fellow and that the sentry's family had been killed in a local bombing. In any case, the driver had been delivering milk to a Krankenhaus (hospital) up on the hill, and he was not quite through with his deliveries yet. He asked if we would mind if he finished his rounds before taking us back to our POW column, and we were more than happy to oblige! He then pulled out a hip flask of Schnapps, had a drink, and offered us one, which we were most happy to have after our extremely sobering past few moments with the sentry. This finished off the bottle, so he stopped to get another. We all had another drink. After perhaps another thirty minutes or so he was through with his deliveries and stopped to inquire by telephone where the POW column had gone to. He couldn't find out, so he said he would take us to a nearby Luftwaffe training base on the outskirts of town—which he proceeded to do.

This base was of beautiful permanent structures, well laid out and of modern brick buildings. He took us in and shortly he was relieved of us. We thanked him and bid him adieu. A German Luftwaffe major took charge of us. First he deposited us with another officer at the officer's mess. The officers were then having tea about 4 to 4:30 P.M. They served us also, while the major attempted by phone to discover where the POW column would be encamped for the night. About 5:15 P.M. or so, he had found out and said he would be returning us. Shortly, a large shiny black Mercedes Benz open touring car flying Nazi and Luftwaffe flags pulled up in front of the officer's mess with a chauffeur driving.

The major told us this was our transportation and asked us to get in. He got in front with the driver. We three sat in the large back seat. There were two jump seats unused. As we were driving out of the base, five or six German Luftwaffe girls in uniform were thumbing a ride into town. The major wanted to know if we would mind if he gave the girls a ride. We said, "certainly not—please do." They got in and used the jump seats and the others sat upon our laps—one each!

They seemed to think this was amusing and laughed and giggled. The three of us thought it was great, too, and actually managed to hold them perhaps a little tighter than would normally be necessary.

Shortly, they were let out and we proceeded eastward out of town. It was now just about dusk, but the sun had not quite set yet. We were traveling down a dusty gravel road, leaving quite a cloud of dust behind us, although we were not going over 25 to 30 mph. There ahead of us off to the left in a large stubble field was the POW column encamped for the night—perhaps 500 to 700 POWs plus the guards. The POWs were spread out all over the field in small groups—all with their little campfires burning, cooking their evening meal.

As we approached in the car, all eyes were turned towards the large shiny black Mercedes Benz touring car which had the Nazi and Luftwaffe flags flying. The car stopped, the major got out, and we three Yanks got out of the rear. The senior British officer (a group captain) and his adjutant (a wing commander) walked over, saluted the German major beside us, and the major very properly clicked his heels and saluted them back. A few words were spoken between them. The major then turned to us, saluted us, and wished us good luck. We, in turn, saluted him, and thanked him for his courtesy, refreshments, and for returning us to the column. He nodded, turned to his driver, and was off. All eyes of the POW group had been upon us as this transpired. The looks on the faces of our brother POWs were priceless and unforgettable.

After crossing the Elbe near Hamburg, the POW column slowly continued its migration eastward toward Lubeck. About April 1, Fessler and his comrades were quartered in a large estate at Trenthorst, a few miles from Lubeck. Most of the POWs were quartered in a large barn, but Nichols and Fessler occupied the loft in a nearby farmhouse. There, Fessler said, they waited out the war.

> We simply lived there going and coming as we wished. Daily, we would go to the other POWs in the barn to learn the latest news, get our share of any rations, etc. Our POW column had a secret radio receiver and actually also a crude code transmitter available any time power was available to us. So we knew pretty exactly the military situation, and our people knew of our location. All knew it was a question of weeks before the final collapse of Germany.

Yet another group of Eagles previously caged at Sagan spent February and March at Stalag XIIID in the Nurnberg area, before one last push to Moosburg.

Bill Hall and others in the sick bay at Stalag Luft III were loaded onto boxcars on February 6, 1945. He counted 38 POWs and eight German guards. The cars were crowded, and the prisoners had to try to sleep standing up. On the first day, there was an air raid. The Eagles held their breath—the train was loaded with military supplies and was therefore a legitimate bombing target. Fortunately, their voyage continued, although they had to endure five days of extreme discomfort. Brewster Morgan recalled his approach to Nurnberg in a December 1977 article in the *Honolulu Advertiser*:

> As I had been wounded in the leg and was unable to walk with the majority of the prisoners, I was moved with hospital cases in boxcars to within 25 miles of Nurnberg. We were unable to approach the city on the

railroad line because of heavy bombing day and night, so we walked the remaining distance. The Germans guarding us on the march indicated we soon would be at a very fine camp with lots of food. It was rather difficult not to feel a sense of anticipation of better conditions, and of eventually getting home to my family, as it was obvious that the German military situation was deteriorating badly.

Morgan found out quickly that the camp at Nurnberg was far from what he had expected:

> It was built as a holding station for Jewish prisoners on the way to the gas chambers. There were no beds except sloping shelves, no bedding except ticks filled with straw.
>
> The food situation was even worse, a daily ration of green soup made from cabbages without any protein. The camp was soon rampant with intestinal pains and other difficulties. Water pipes had frozen. There was no heat of any kind until we tore down the shower houses to stoke our stoves.
>
> We remained in this camp until the first part of April, under near-starvation conditions. One day we saw at the gates a line of Red Cross trucks driven by Swiss. Each man was issued a week's food supply—one parcel—which we proceeded to consume, to the detriment of our system. The next week we were too sick to eat anything.
>
> The bombing of Nurnberg continued day and night. We left our barracks for slit trenches because of the pieces of unexploded shells that descended from the sky during raids. I recall a duel between a British bomber and a German fighter. The exchange of machine gun tracers could be seen clearly as the huge bomber passed our camp, caught fire, and crashed into a nearby forest. The fighter, wounded, too, followed the bomber over the camp and crashed almost into the funeral pyre of the bomber.

Bill Hall's diary entries between February 11 and April 3, 1945, vividly captured the hunger, cold, sickness, and aerial devastation afflicting the Nurnberg contingent:

> Feb. 11 Left the sidedoor pullman around 10 A.M. Marched to the Nurnberg camp and found men from the Stalag Luft III West, South and North Camps there. Very short of food. Sleeping on boards again but plenty of room. Bugs are plentiful.
>
> (Later) This Nurnberg Camp had been a Jewish concentration camp. We slept on a board platform about two feet off the ground, six feet wide and 40 feet long. Slept with our clothes on but were bitten all night by bedbugs, fleas, and lice.
>
> Feb. 13 Still hardness on boards. Slept with Stanhope as I got caught out in an air raid.
>
> Feb. 14 Three appells (roll calls), finished up with an air raid. Flak and one pill pretty close. Moved back with RAF.
>
> Feb. 16 Back to two appells. No coal for cooking. Lots of wood scrounging. As yet no air raid.
>
> Feb. 17, Sat. Same damn routine. Rumors of Switzerland flying fast. Still crowded as hell, sleeping 24 to the bay six inches apart.

Feb. 20 Air raid by U.S. Army Air Corps; lasted for two hours. Could see the bombs falling on Nurnberg along with five kites (planes shot down). Sweating out RAF raid for tonight.

(Later) No RAF, but very heavy explosions.

Feb. 21 Another air raid on Nurnberg. The pills were closer. Had to hit the deck a few times. No lights.

Feb. 22 Another raid hitting town close by. Went out for a parole walk, as far as Nurnberg Ring. Several bombs had hit.

Feb. 23 Raid some distance away. Saw two P-51s stooging by with flak bursting close to the rear.

Feb. 24, Sat. Food getting scarce. Bad rumors about the food in Senior Officers' mess.

Feb. 26 Spud ration cut. Rumors flying fast about S.O. mess concerning good eats. Air raid again. Mosquito or heavy bomb landed extra close to camp.

Feb. 27 Air raid on again.

Feb. 28 Three air raids.

March 8 Hot brew (water with coffee rushed through) and one thin slice of bread (one-seventh of a loaf per day). Dinner of Spam (one tin between six men) and spuds, three small ones. Soup at 11 A.M. and P.M. (one cup per person). Another brew after 5 P.M.

March 9 Rumors that rations to be cut again 45 percent.

March 12-14 Last of Red Cross rations. German rations going rough. No energy to do anything with.

March 15 Half issue of Red Cross parcels.

March 16 RAF raid on Nurnberg at night, large fires started, bombing very accurate. Pathfinders came in and marked our camp, which is in the middle of a train marshaling yard, before the bombers hit.

March 20 Our turkey dinner; dry vermouth, cream of tomato soup, 12 lb. turkey with dressing, ice cream with hot mince pie. (Only a dream.)

March 23 S.A.O. is taking Red Cross parcels from the officers; disciplinary actions for pulling lumber off buildings to burn.

March 25 12 carloads of Red Cross parcels in, and 12 P-38s came across the sky.

April 2 Eight P-38s stooging past. Didn't feel well, went and laid down after dinner. Pop Coventry took my temperature; was sent to "Riviera," our hospital. Double pneumonia and dysentery.

April 3 Sick all day long. Hospital evacuated at 5 P.M. They left 23 of us behind as too sick to move.

On April 4, Hall noted, the main camp of POWs at Nurnberg moved out. American troops were driving on Nurnberg from the north and the southwest, and the Germans intended to keep the POWs from being liberated there. The kriegies' destination was Moosburg, but for such Eagles as Bill Geiger and Brew Morgan, the 10-day march of over 80 miles provided some perilous opportunities to escape. Bill Geiger recalled one unplanned deviation:

> My walking companion on this march was a Scotsman, Archie Galloway, a tremendous little guy. He and I were walking along one night

in a column of some 10,000 POWs, in the rain with our heads down. At one point Archie nudged me and I looked up and we looked around and couldn't see anybody at all. Archie turned to me and said, "Son, I think we've escaped."

The column had turned right or left somewhere along the line and we, walking along with our heads down, had just kept on going straight. We were only gone for two or three days. It was a very dangerous place and a very dangerous time to be off on your own, and eventually we ran into the column again—somewhat glad to do so, I might add.

Brew Morgan and a friend experienced a wide spectrum of emotion during the week they left Nurnberg:

We moved out on a beautiful clear morning and looked down on Nurnberg, totally in flames with hardly a building standing. I wept inside for the civilians caught in the terrible crossfire between the German and American armies. We marched all day and into the night, stopping each hour for 10 minutes. It started to rain very heavily, and we staggered in a column of four, accompanied by exhausted dogs and guards.

One morning, my closest friend woke me up and said it was time to escape. His plan was to leave the marching column, head west, and attempt to reach the Americans who were only 50 miles away. After four or five days very near the front, we were captured by Germans. Too busy to accompany us, they told us to return to our marching column. Chilled by the prospect of attempting to cross the battle lines, we rejoined the column, which finally arrived at the massive camp Stalag VIIA, which was quite near Dachau, the Jewish camp.

And so, having endured incredible hardships during the early winter months of 1945, the caged Eagles of Sagan were dispersed in Moosburg, Nurnberg, and Trenthorst. Like their compatriots in Barth, they anxiously awaited their respective moments of liberation.

13

Liberation

AS THE BRUTAL WINTER turned into the gentle spring of April 1945, the Eagles' torment was drawing to a close. Daily the Allied air and ground assault pressed on. Nazi Germany's collapse was near. Their spirits lifted by news of the advance of the Americans and the Russians, the caged Eagles eagerly awaited the exhilarating moment of liberation.

First to experience the taste of freedom was Charles Cook, who had been so ill at Muskau that his friends had despaired for his survival. Saved from death by the miracle drug penicillin and carefully attended by German doctors, Cook was among a small number of patients in the Sagan hospital. Suddenly, shots rang out, the German guards fled, and the Russian army came in.

> The Russians gave us food and moved us to a hospital in Rawicz, Poland. Two British Indian soldiers died in the van I was in.

Soon, the group was loaded in boxcars for rail shipment to Odessa on the Black Sea.

> As a first lieutenant, I was placed in charge of 65 British and American Army and Air Force personnel. We also had six American soldiers who had lost their minds on the G.I. front lines to watch over. I could hardly walk, much less give orders.
>
> The British medical orderlies with us were wonderful. They told me, "Don't you worry; we'll take care of things." And they did. It wasn't easy traveling, but at least we had shelter. Up ahead of us there were a lot of Russian civilian men and women in open cars, really suffering from the cold.

Cook spent a week at Odessa, recovering slowly, and then boarded the S.S. *Madeira*, which was carrying 3,000 British and American war prisoners to Naples, Italy.

> We had lots of lovely food but I couldn't eat much. "The British medics did a beautiful job. When I arrived at Naples they had an ambulance and nurses waiting to take us to a hospital.
> Two other pilots who had been shot down said they had enough of flying. They took a freighter home. I flew to Casablanca in a C-47, then caught a C-54, the first I had seen. It was supposed to go to Stephenville, Newfoundland, but I was having a relapse and they dropped me off in the Azores. I spent three days in the hospital there, then went on to Washington.

While Cook was embarking on his journey of freedom, American forces were coming closer and closer to Nurnberg. Bill Hall, who had been too sick to join the column of prisoners moving out to Moosburg on April 4, chronicled the last two weeks of his captivity in his diary. His entries show dramatically the fear and triumph of those days:

> April 5 Quite a bit better. U.S. heavies came over and hit Nurnberg railyards; 28 Americans killed coming from the station.
> April 7 Had a wonderful shower, my second at Nurnberg. Later, at the Nurnberg Camp, 50 of us at a time were taken to a room that resembled a walk-in frig. It had a thick, air-tight door, ceiling about seven feet high with the shower jets protruding, ceiling and walls of solid wood paneling. There was an odor of death about the place and as we walked in we thought we were in for it as well. An excuse was made, and several fellows stayed in the doorway to prevent the door from being closed. The German general in charge of POWs inspected the hospital, ordered able-bodied POWs be ready to evacuate.
> April 8 Yanks came over, started to drop the old pills, kept it up for about an hour on all cardinal points around Nurnberg.
> April 9 Four P-47s shooting the area up, doing a good job.
> April 11 USAAF all around Nurnberg, then RAF appeared and blasted hell out of Nurnberg, using Darby Lanks.
> April 12 Bread very scarce. Four-kilo loaf sells for one tin of corned beef and 10 cigarettes. Sardines selling for 10 to 15 cigarettes. Beans two packs of cigarettes per bowl.
> April 14 Plenty to eat; another Red Cross parcel issued again. Yanks getting closer; heard artillery in the distance.
> April 15 P-47s out in force. Germans evacuating rest of camp; typhus placards put up on main gates. Our troops reported on three sides of Nurnberg. We're digging up new shelters.
> April 16 Artillery opened up on Nurnberg about 7 P.M., plenty heavy all night long. Spent most of night in trenches. A few shells landed close. There also were a number of air bursts directly over the Italian kitchen about 25 yards from our shelter. Germans are blowing up their ammo dumps.
> April 17 Our day of liberation. I was giving out the sugar ration when someone yelled, "Tanks coming into the compound." I ran to the window.

Sure enough, there they were—American tanks with American infantry, the 45th Infantry of the Seventh Army. I was so happy, tears of joy just ran down my cheeks. I had no control over them. Then I saw the goons run up white flags. I was going around in circles! The field ambulance pulled in, too, and loaded us down with brandy. The Germans guarding the camp gave up without a struggle. I got very drunk. Another heavy artillery barrage afternoon.

April 18 Still walking around in a daze; haven't yet realized that we are liberated. The Americans are coming in and going out. Got a good souvenir. After the field ambulances passed through, General Omar Bradley showed up in a Jeep. The back of it was loaded with cases of German schnapps or brandy, and his driver was handing this out. Bradley talked for a while and then handed me a Colt 45 and said, "Take care of yourself." As he left the Germans were shelling the area. I was sitting on the pavement in front of the sick bay and picked up a hot piece of shrapnel that landed beside me. One of the fellows got me into a slit trench.

A Sgt. Holman from Arkansas, from the 45th, said he would get me a German P38 semi-automatic pistol, which I wanted very badly. He could speak Russian. The Russians were in a compound close by, so the American armed us with a hand grenade each and told me if there was any trouble to pull the pin fast. We walked into the Russian barracks, and those fellows were in as bad shape as anyone could possibly be. There had been no food for them. We walked through the barracks and up to a small back room. The Sergeant asked the Russian for a P38. The Russian went away and came back with a brand new P38. I still have both guns.

April 19 Loaded onto a truck and started out of camp. Moved around in a large semi-circle through Bamberg where I pretty near got shot. At Wurzburg the town had been taken down to the ground. Going through the village we saw only three civilians. It did the old heart good.

April 20 Moved from 11th Evacuation up to Wurzburg airdrome. Went out to one of the P-47s and sat in the cockpit. Did it feel good! Sat around and talked with the nurses. It was wonderful to hear their voices again. Was called at 4:30 P.M. to leave for England, but didn't take off until 6:50. Then we almost copped it from another C-47, which nearly ran into us on takeoff. Trip was successful except for a couple of instances of severe icing conditions. Also, the hydraulic system blew out on us when the pilot went to turn on George. The crew was successful in holding the pressure till we landed in England at 11 P.M., at a drome near Oxford.

Hall and the other patients were placed in an American evacuation hospital near Oxford initially, and then were sent first to an RAF hospital at Swindon, and finally to the RAF hospital at Cosford.

We were supposed to stay in the hospital to recuperate and fatten up. Instead, two of us went out the window of our hospital room and took a train for Scotland. After three or four days, I hitched a ride on a B-17 back to London and made my headquarters at the Jules Club.

Shortly after we returned to England, we ex-POWs were visited at the Club by Queen Elizabeth and by Princess Elizabeth dressed in her Army

Transport driver's uniform. We all lined up, and the Queen, followed by the Princess, moved down the line and chatted and shook hands with us.

Now, many years later, I have no idea what motivated me, but I do remember asking the Princess to have dinner with me. And I clearly remember her kind and very diplomatic reply. She said to me, "I am terribly sorry but I have a previous engagement with my mother this evening."

Tall, dark-haired and well built, charming, and not at all shy, Hall was regarded by some of the London women who later married Americans as "just about the handsomest of all those good-looking Eagles."

I was in London for the month of May and was there on V.E. day when all London went mad. The RAF shipped me back to Washington on a U.S. Transport Command C-46 on June 7.

* * * *

During the last three weeks of April, the prisoners at Moosburg, like their comrades at Nurnberg earlier in the month, were confronted with numerous air raids. George Sperry described the situation:

> At no time during the day or night was the air above our section free of Allied aircraft. Inside fences of our camp were broken down, and we could move freely from one compound to another.
>
> The push of the American Seventh and Third armies into western Germany precipitated many rumors. On our maps we could watch the advance toward us. It appeared that we would be liberated before long. Every morning, FW 190 fighters could be seen hedgehopping over the camp as they attempted to escape observation from the higher-flying American fighters that were becoming more numerous everyday. There were heavy air raids on the cities around us. We could see B-26 Martin Marauders and other medium bombers attacking targets on all sides. At last, on April 26 and 27, artillery rocked the countryside. We could hear the 105s whistling over our heads as we lay stretched out on our sacks.

Barry Mahon, who only weeks before had returned to Moosburg, reported that on the evening of April 28 General George Patton, with his staff, came to within sight of the camp, met with the SS commander, and gruffly put the Germans on notice:

> Patton had 12 tanks and three jeeps, and his whole Third Army was at least a day behind. But the SS, the only ones still fighting, didn't know that. Patton called for our colonel and some of the representatives from the camp to meet with him and the SS commander and decide what was going to happen with the prisoners.
>
> The SS commander said he would allow the camp to be captured and bypassed without harm if Patton came in with no offensive weapons. Tanks are offensive weapons, so there was no way he possibly could stop the offensive. Patton just said, "Colonel, if any of my men in that camp are

hurt or killed because of you, it's going to be your ass. So you just act accordingly."

Convinced of the American intention to liberate Moosburg immediately, the Germans pulled most of their troops out by truck during the night, leaving only a small force to guard the camp. George Sperry recalled the vivid images of the next day:

> The morning of April 29, 1945, dawned clear, without wind or cloud, and there was a feeling of expectancy among the several thousand of us kriegies. For the first time in many months we arose from our straw pallets with a different feeling about the new day. Rumors were flying like wildfire all over the camp as the sun rose; American armored forces were only a few miles away. German resistance was stiffening. We were to be moved to the mountains toward the east, to be used as hostages as a last-ditch resort. SS troops were moving into the immediate area to execute all Allied airmen in one last blood bath.
>
> Gradually, that dramatic early morning, our situation changed. By 8 A.M. the Luftwaffe and army guards had deserted their posts and joined us kriegies within the fence, turning their arms over to their former prisoners. Meanwhile, SS troops had surrounded the camp, moving anti-tank guns into place on the main road in front of the establishment. This was it, we told ourselves.
>
> But around 9 o'clock we heard the tanks of the American 14th armored division, and soon we could see them approaching our barbed-wire resort. Kriegies climbed on top of buildings and guard towers to watch the excitement. Bullets ricocheted over the compound. A couple of kriegies were hit, but none was injured seriously. There was a short, futile battle by the diehard SS troops, and finally we were liberated amidst a terrific outburst of cheering. Feelings not expressed for many long months—in my case, two years, seven months, and three days—found release in noise-making and in good-natured horseplay.
>
> The crowning development on this Eagle morning, this happy day, was not that we were no longer prisoners of war, and that soon we would be going home. In the midst of all this celebration the true climax was that the American flag was rising on a church steeple in the town only a short distance away. At this announcement the American soldiers came to attention and saluted. Among us kriegies, every man felt the trickle of tears falling down his cheeks. Our feelings of pride in country and service overwhelmed us. We were free men once again.

Within days, the Eagles at Trenthorst and Barth were also liberated, by the Canadians and Russians, respectively. The Barth contingent was evacuated in B-17s of the Eighth Air Force to Britain and to Camp Lucky Strike, in Normandy, where as Bob Patterson described it, the ex-POWs were "deloused, given uniforms, fattened up, and made ready for departure to the U.S."

A few of the Eagles, however, found alternative means to celebrate their return to the world outside their cages.

Don "Snuffy" Smith, who took special pride in the fact that Moosburg was liberated by the Third Army's 99th Infantry Division, which had trained at Camp Maxey in his hometown of Paris, Texas, decided after a few days to explore "the other Paris." Smith and his friend, Captain Homer Foister of Boynton, Oklahoma, caught a ride on a mail truck to Nurnberg, then reached Wiesbaden and took a flight to Paris, expecting to garner a great meal. Their plans nearly went awry, as Smith recounted:

> It was Sunday, and when we tried at the finance office to get some of our back pay they would not let us have any money because we had not been processed and identified. Finally we borrowed some money from the Red Cross, went to a restaurant and bought a big steak, ate about two bites, and were full. Our waiter looked delighted. That steak did not go to waste.

M.E. Jackson took himself on a jaunt, first to Paris, and then to England, where he regained some of the strength and weight he had lost in the POW camps.

> I got a motorcycle and rode around Germany for several days. Then I had four or five days in Paris, and then went to England.
> I was picked up by the MPs and placed in hospital quarantine near Cambridge for 16 days. It was very pleasant, and I gained 26 pounds—very good for a six-footer.

Barry Mahon continued his wide-ranging explorations of Europe, starting off from Moosburg in one of Patton's tanks:

> I asked for permission to go to Munich with a friend. This friend had pulled up in one of Patton's tanks and asked if anyone from California were here. I said, "Yes." He asked what town. I told him. What street address? I told him, and he said, "I live next door."
> "No you don't; a little girl lives next door," I said.
> "Yeah, I married her," he replied.
> What a small world! So I got permission to go with the tank commander, and we chased the SS down toward Munich. It was a gay kind of a fight because there wasn't much resistance. As we went through the various towns we would liberate cameras and binoculars, and in one warehouse they had liberated a bunch of ski-troop rabbit fur jackets. So immediately every GI in the Third Army had on a ski jacket.
> When we got to the outskirts of Munich, this friend of mine got me a ride in a jeep to Luxembourg. I was walking down the street there with a four- or five-day growth of beard, wearing an American paratrooper's jacket with RAF wings sewed on, GI pants and shoes. A corporal MP came by, stopped me, and said, "Soldier, you are the worst-looking mess I have ever seen."
> "Okay, take me to headquarters," I said.
> At headquarters they asked for my identification. When they learned I was equivalent to a captain, they were a little more respectful. In fact, the lieutenant took me home to let me shave and clean up at his house.
> He was dubious about my ability to get a flight out of Luxembourg

because only VIPs were flying out of there. But when the shuttle DC-3 or C-47 came in, I went to the pilot and said, "Look, I've been a POW for three years and I'd surely like to get to Paris tonight. How about letting me fly in the cockpit?" He said, "Sure, get on board."

All the congressmen and generals in the waiting room saw me get in. On the way down, a colonel came up to the cockpit to find out who I was. When he found out the story, he said he would take me out that night in Paris, on him. When he landed in Paris we were informed that Eisenhower had just declared the following day to be V.E. Day, and already you could hear the city beginning to celebrate the news.

The colonel was up to his word because he took me out for one of the wildest nights I have ever had. Before going out I checked into a hotel. There was a GI re-outfitting center where they would give you new uniforms. There was no such a thing for the RAF, so I went through this one, receiving combs, brush, toothbrush, new shirts, and so on. Then the other boys were in a pay parade, so I got in line and they handed me $50. The Americans could draw that against their pay. I gave them my RAF serial number, and I guess to this day they're trying to figure out where that $50 went.

That evening the colonel took me out, and I began to realize that Paris was not a cheap place. I ended up at the RAF Officers Club and, having three years' back pay in the bank, was allowed to cash a check for the equivalent of $1,000. This lasted several days until I was poured on a shuttle plane to England and later taken to Prestwick, Scotland, and flown home to Washington, D.C.

Morris Fessler and Bill Nichols (whom Fessler called "Nick") also demonstrated ingenuity and a zest for adventure, as they departed their quarters at Trenthorst in northern Germany. Having first obtained some of the weapons discarded by the Germans—a couple of M98 Mauser rifles, a Luger pistol, and a small Belgian Browning Automatic—they proceeded to commandeer a German car and to enjoy the pleasures so long denied them. Fessler recalled those blissful first days of freedom:

> I first picked up a 1935 Ford Sedan, but since it didn't run too well, I left it and then picked up an Auto Union sedan with a 2-stroke engine. It only needed a new spark coil to run properly and thereafter it ran beautifully.
>
> When we first got the car going, we and some of the other POWs who had acquired wheels drove up to Lubeck to the brewery there, drank all the beer we could, loaded up our vehicles with all they would hold, then proceeded back to our fellow POWs for a little celebration.
>
> Within two days Nick and I were ready to move out. We each had acquired a running car; we also acquired about a half dozen 5-gallon petrol cans each from the Canadian Tank Supply Corps, plus some oil. We had obtained new clothing, fresh battle dress, underwear, shoes, etc., from another nearby Canadian Supply Corps group.
>
> The authorities who were in charge of seeing to liberated POW groups and getting them back to England didn't interest us much. We registered with them, kept on the fringes for a couple of days, and then said to hell with it. They wanted to keep us all together like a herd of sheep—like we

had been for the past three and a half years. They moved the POW group a couple of places, then had them waiting for DC-3 (Dakota) flights back to England. We didn't want to go back to England yet. We had just been liberated after all those years in "Hoosegow." We wanted to live a little first!

We debated destinations to visit, like Paris or Marseille, and picked Brussels as the easiest and most practical. Our plan was to take the cars back to Brussels, sell them in the black market there, and live on the proceeds until the money was gone before going back to England.

The Canadian Supply Corps who reequipped us with the new clothing had advised us that when we got stopped at the roadblocks, the sentries would try to take our cars away from us. We were told to demand to be taken to the commanding officer of the area in each case, tell our story, and get a pass through that area. Otherwise, the cars would be confiscated. Also, at night especially, we were told to garage the cars because if the MPs found one unattended, they would confiscate it. They gave us an address in Brussels for a black market contact.

We painted large white stars on the hoods and on the sides of our automobiles and started out, one following the other. We hadn't gone far when we came upon three or four of our comrade POWs whose car had broken down completely. They sweet-talked us out of one of our cars. Nick gave up his car and climbed in with me in the Auto Union sedan.

Shortly thereafter, we came upon a Swiss Red Cross group of four or five trucks who were attempting to locate our main POW group to furnish them with Red Cross parcels, clothing, and blankets. We gave them directions, but first relieved them of as many Red Cross parcels and blankets as our car would hold. The rear of the car was packed full of them.

The roads and highways heading back towards Holland and Belgium had many trucks and vehicles piled full of loot, even tank trailers carrying large boats and yachts—all heading towards the black markets. It really looked funny seeing a tank trailer hauling a 30 to 40 foot yacht from Germany towards Holland, probably for some rather low-ranking officer of the Supply Corps group!

Occasionally, in large open fields en route, we would see thousands of German troops who had surrendered and were being held together pending organized transport.

We proceeded towards Holland and, as we passed from roadblock to roadblock, the sentries would try to confiscate our car; we would demand to be taken to see the commanding officer; and in each case he gave us a pass after listening to our story and usually fed us, too. It worked beautifully!

On May 7, about 6 P.M., we passed out of Germany into Holland. We decided to stop in the first town and spend the night. We went to a local pub, had dinner and drinks, and joined a celebration in progress for the war being officially over "in Holland." We spent the night upstairs with two girls who were at the place. The next morning, after breakfast, we took the two girls home—back across the border into Germany a few miles, where they lived. We were a bit apprehensive about going back into Germany, but nothing happened. Before noon, we were again on our way to Brussels. It was May 8, and we heard over the local radio that the war was officially over; Germany had surrendered. It was also my birthday. I couldn't have had a better gift.

We arrived in Brussels the next day at approximately noon. We immediately contacted these black market people, and by about 3:30 to 4:00 P.M. that afternoon we had sold the car, the Red Cross parcels, and blankets to an auto garage owner, and we had the cash in hand—as I recall, 32,000 Belgian Francs. Nick and I split it right down the middle.

We set off for the downtown area and checked in at one of the downtown hotels. We intended to enjoy every minute of our stay until every franc was gone. I saw a gal in the bar lounge late that afternoon who looked good to me, but she was with another fellow. Through the manager, I sent her a note telling her I would like to see her if possible. Later that evening, I met her and we got together. We hit it off real well. I appreciated her and she seemed to like me. She could not spend the whole night, but had to get up and go home. But the next day I would see her again about 3 to 4 in the afternoon until after midnight. Saturday and Sunday, we spent the whole day together. We had some meals sent to the room, had others out. We lived high on the hog with good wine and champagne and lots of loving.

I would generally have breakfast and lunch with Nick, and we would talk some, but other than that we didn't see each other, since we went our separate ways and we both had serious work to do to catch up for the past three and a half years.

After seven days, our money was about gone. Nick and I went out to Brussels airport, registered with the RAF authorities there, and climbed aboard a DC-3 for our flight back to England. We landed back in England on the afternoon of May 14, 1945.

After a couple of days of interrogation, hospital testing, and outfitting common to all POW returnees, I had three months' leave back home in the States.

Two Eagles stayed behind briefly in Germany while their comrades found refuge and adventure elsewhere.

Ironically, Bill Geiger, who had suffered from hunger in several camps and endured the "green-death" soup at Nurnberg (which he tersely called "a vegetable soup for cattle"), was in charge of most of the food in Moosburg for a while:

> I was given command of a guard detail guarding a factory. I am not sure why I got this particular post, but I gathered it was because I was one of the older POWs, and they were trying to give us something to do while we were waiting to go home.
>
> This factory held all the remaining Red Cross food around. There was nobody, of course, to guard that food from me, so I ended up by almost owning Moosburg. When it came time for me to leave I volunteered my services for another month or so. They were refused, and I found myself on my way home.

Up north in Barth, Bill Edwards stayed on as a clothing and food officer for the Russians who had liberated the camp:

I worked with the Russians on such projects as locating farms that had cattle. The Russians did not have supply lines as we did. They were living off the land, and we needed cattle to slaughter for beef.

There were several Russian officers—a major, a captain, and a couple of lieutenants—in the camp with us, and we worked with them to solve our supply problems. Some of them were used to pull some supply wagons when we didn't have other transportation, and this led to trouble. The Russians decided that I had contributed to atrocities against the Soviet government by using Soviet forces, including some officers, to pull wagons. They said they were going to court-martial me, but then one day, out of a clear blue sky, they told me to go home. Obviously they decided to get rid of all Yanks.

* * * *

And so, the caged Eagles were free men once again, in a world they and their comrades had saved from tyranny. Many would pursue a career in military service or commercial aviation. Yet other former Eagles would return to business pursuits across America. Whatever their future course, these heroes of battle and survivors of hardship had, by their courage, helped preserve the spirit of freedom for future generations.

Index

1st Squadron, RAF, 16, 75
111th Squadron, 141
121st Eagle Squadron, 16, 40, 52, 56, 71, 72, 76, 87, 108, 117, 119, 134, 141
131st Squadron, 141
133rd Eagle Squadron, 24, 25, 52, 76, 81, 82, 87, 88, 111, 128, 132

242nd Squadron, 43
258th New Zealand Squadron, 32, 43-45

31st Fighter Group, USAAF, 96
324th Wing, 141
334th Squadron, 11, 87, 105
335th Eagle Squadron, 87, 107, 117, 119, 120, 141, 142
336th Eagle Squadron, 83, 87, 128, 129, 132
353rd Fighter Squadron, 143, 148
354th Fighter Group, 143
359th Fighter Group, 142

4th Fighter Group, USAAF, 11, 105, 108, 120, 127, 132, 141, 151

56th Squadron, 75

605th Squadron, 43
609th Squadron, 7

71st Eagle Squadron, 8, 11, 13, 15-17, 21, 32, 56, 72, 75-76, 87, 105, 119, 140, 143, 151
78th Fighter Group, 25

99th Bomber Squadron, 42

A
Abbeville raid, 72, 73
Abrassard, Albert, 146-147
Abrassard, Madelaine, 146
Achilles, Theo, 7
Adams, Vernon D., 164
African campaign, 44-45, 52-70, 141

Alexander, Richard L. "Dixie", 120-127
Allen, Luke, 9
American Eagle Club, London, 35
American Magazine, The, 68
Anderson, Stan, 105
Armée de l'Air, France, 4
Astaire, Adele, Lady Cavendish, 40-42
Athenia shipwreck, 127-128
Auswertestelle West interrogation center, 77
Ayres, Hank, 107-108

B
Bader, Douglas, 18-19, 23, 96
Baker, Bill, 83
Barcelona, Spain, 75, 93, 94
Barclay, George, 30
Batavia, 48
Bateman, Charles, 9
Bath, 181
Battle of Britain, 1-3, 7-8
Baxley, Ed, 139-140
Beaty, Dick, 81
Beeson, Duane, 120
Belgium underground, 144-148
Berlin bombing raids, 118, 120, 129, 131-132
Bethune raid, 22
Bicksler, Edwin "Bix", 52-55
Bienfait, Yvonne, 159
Biggin Hill, 75
Blakeslee, Don, 120, 128-129, 134
Blitz, London (*see* Britain, Battle of Britain)
Blitzkrieg on Poland, 1
Bodding, Carl O., 72
Boersma, 114
Bolitho, Hector, 10
Boock, Robert, 105
Boyles, Frank R., 76, 120
Bremen raid, 113-117
Brest raid, 5, 81, 83, 111
Brettell, Edward G., 81, 84
Brettell, Gordon, 111

Brewster aircraft, 9
Britain
 Battle of Britain, 1-3, 7-8
 civilian aid to Eagles Squadron members, 35-42
Brown, George, 9
Brussels raid, 108, 119
Button, Harry, 30

C

Cairo, Egypt, 52-54, 56, 69
Campbell, John "Red", 32, 43-52
Care, Bud, 106, 120
Carpenter, George, 119-120
Cavendish, Charles, 42
Chagny, 74
Chap, Norman, 52, 55
Childers, James S., 13
Church Fenton, 8, 10
Churchill, Walter M., 8, 72
Churchill, Winston, 1
Clanan, Thomas, 103-104
Clark, Albert P., 96
Clark, James A. Jr. "Bud", 105, 109, 120
Clayton Knight Committee, 15, 35
Cleary, 63, 64, 65
Coen, Oscar, 26-33, 43, 75-76, 80, 105
Cologne raid, 20
Columbo raid, 44
Cook, Charles A. "Cookie", 81-84, 86-88, 96, 108-110, 126-127, 134, 139, 140, 161-164, 177-178
Cooper, Duff, 12
Corson, 63, 64

D

Daley, Jim, 72
Daniel, Gilmore "Danny", 22-24, 71, 108
Daniel, Pete, 7
Day, Harry M.A., 107
Daymond, Gregory "Gus", 9, 11, 17, 27
de Valera, Eamon, 40
De Vroom, Maurice, 145-146
Debden Airfield, 76, 148, 151
Denmark, German invasion of, 1
Dieppe raid, 71, 75, 76
Dobbyn, Harry, 47, 48
Donahue, Arthur "Art", 9, 32, 43-45, 47, 50, 52
Doorly, Eric, 91-95
Douglas, Sholto, 87

du Maurier, Daphne, 41
Duff-Smith, Michael Assheton, 40
Duke-Woolley, R.M.B., 105
Dulag Luft POW Center, Germany, 77, 84, 96, 107, 116, 141
Dunkirk evacuation, 5, 8, 79
Dunn, William R., 17, 27
Dutch underground, 114, 155-160

E

Eagle Club, London, 50
Eagle Squadrons
 formation and history, vii-viii, 3, 8
 recruitment, Clayton Knight Committee, 15, 35
 transfer to U.S. Army Air Force, 80-81, 87
 uniform for pilots, crews, 8
Edinburgh, Scotland, 36-42
Edner, Selden, 72, 76, 120
Edwards, Bill, 186
Edwards, Wilson V. "Bill", 132-134
Eiben, Fritz, 62
Eichar, Grant, 76
El Alamein, Battle of, 57-68
Ellington, Paul "Duke", 117-120
Emden raid, 105
Emerson, Don, 129
Emmerich raid, 107
Evans, Roy W., 141-142
execution of escaping POWs, 111-112

F

Fardel, Joseph, 30
Fenlaw, Hillard, 14, 15
Ferris, 117
Fessler, Morris "Jack" 15, 23, 103-104, 108, 169-173, 183-185
Fetrow, Gene, 75, 105
Figueras, Spain, 31-32
Fink, Frank, 108, 138-139
Finland, German invasion of, 2
Finnegan, 63, 65
Flying Tigers, 3, 56
Foister, Homer, 182
France
 Armée de l'Air, 4
 French Underground (*see* Free French Underground)
 German invasion of, 1, 5-6
 Lafayette Escadrille, 3, 8
 liberation of, 14, 15, 21, 154

Index 189

France, Vic, 120, 151
Frankfurt air raids, 143
Frankfurt am Main interrogation center, 19-22, 77, 84, 87, 119, 133
Free As A Running Fox, 103
Free French Underground, 19, 30-32, 74, 75, 90, 91, 93, 138-139, 149-154, 159, 167

G
Gabreski, Francis, 138
Galloway, Archie, 175-176
Gaunt, Jeffrey, 8
Geffene, Don, 32, 43-44, 53
Geiger, Bill, 21-23, 71, 79-80, 95-96, 99, 104, 108-109, 162, 175, 185
Gentile, Don, 120
Gerard, Pat O'Leary, 74
Gibraltar, 32, 43, 44, 94
Goodrich, C.G., 162
Goodson, Jim, 120, 127-132
Gray, Jim, 140-141
Greece, Greek underground, 62-69
Griffin, James, 52-56
Gutersloh raid, 154

H
Hall, Don, 53, 54
Hall, William I. "Bill/Kriegie", 16-21, 23, 27, 71, 77, 79, 96, 97, 99, 100, 108-109, 140, 162, 173-175, 178-180
Happel, Jim, 118, 120
Harp, Carter, 76
Harrison, Tommy, 18, 19
Hildburghausen POW hospital, 20-21, 77
Hitchcock, Tommy, 134
Hitler, Adolf, 1-2
Hively, Howard, 120
Holmark hospital, Belgium, 20
Hubbard, Thomas H., 159
Hughes-Stanton, Blair "Cappy", 78
Hurricane aircraft, 10, 44, 45
Hurricane Over the Jungle, 46

I
Ireland's role during war, 24-25, 40-42

J
Jackson, Marion E. "Jack", 81, 84, 96, 98-100, 112, 165, 182

Japanese, War in the Pacific, 45-51
Java Island, 47-49
Jenkins, 40
Jones, William L.C. "Casey", 40-42, 71-72
Jubilee Operation, 75, 77

K
Kelly, Frank, 7
Kelly, Mike, 56
Kelly, Terence, 46
Kelly, William P., 76
Kennard, Hugh, 76
Kennedy, William L., 99
Kennerly, Byron, 9
Keough, Vernon C. "Shorty", 4-5, 7, 8, 13, 15
Kepner, William, 128
King, Coburn, 76
Kirton-in-Lindsey airfield, 10, 56, 72
Kluckholm, Frank L., 80
Knickerbocker, Bob, 11
Knight, Clayton (see Clayton Knight Committee; Eagles Squadron)
Kolendorski, Stanley, 9, 11

L
Lafayette Escadrille, WW I France, 3, 8
Last Flight from Singapore, 47
Le Bourget raid, 108
Leckrone, Philip "Zeke", 9, 12, 15
Leipzig air raid, 164
Lewis, John D., 109
Liberty Magazine, 9, 13
Lille Hospital, 23
Lille raid, 16
Lincoln Brigade, Spanish Civil War, 2
Livers, Lennie Marting, 70
London Blitz, 2, 7
London Can Take It, 10
London, Diary, A, 9
Low, Bob, 9-11, 13
Lubeck, Germany, 169

M
MacFarlane, Lee, 105
Madrid, Spain, 74
Magisters aircraft, 39
Mahon, Jackson B. "Barry", 76-77, 98, 100-103, 111, 166-169, 180, 182-183
Malta, 32, 43

190   Caged Eagles

Mamedoff, Andrew, 4, 7, 8, 11-13
Mannix, Bob, 17
Maquis (see Free French Underground)
Maranz, Nat, 15, 16, 71, 108, 169-173
Marseille, France, 31
Martin, Donald, 117
Marting, Harold F., "Hal", 56-68, 69-70
Martlesham Heath, 13, 16, 17
Martonmere (see Robinson, J. Roland "Robbie")
Matthews, Joseph, 108
Mauriello, Sam, 75
McCall, Hugh, 13
McCarthy, 20
McColpin, Carroll W., 82
McGerty, Thomas, 21
McGinnis, James, 9
McMinn, 63
McNamara, J. J., 134
McNickle, 110
McPharlin, Michael G. "Wee Mac", 75-76
Melidi, Angela A., 68
Metro Goldwyn Mayer Studio Club News, 13
Middleton, Drew, 11
Middleton, George, 88-90, 98, 100
Mills, Hank, 119-120
Miluck, Mike Eddie, 56, 69-70
Miranda de Ebro prison, 93, 94
Montoisey, Marcel, 146-147
Mooney, Jack, 76
Moore, Arthur "Jim", 9
Moosburg, Bavaria (see Stalag VIIA)
Morgan, Brewster, 105-107, 173-176
Morlaix raid, 81-94, 96, 111, 128
Munich raid, 132-133

N

Nee, Donald, 105
Neville, Gene, 81, 83, 84, 86
Nichols, Bill "Nick", 15, 16, 23, 104, 108, 169-173, 183-185
Nimes, France, 31
Nitelet, Alex, 30
Niven, David, 10
Normandy invasion, 113
North Weald airfield, 13, 15, 56
Norway, German invasion of, 1
Nurnberg POW camp, 173-176, 178-180

O

O'Berg, 63, 64, 65
Oberursel, 132-134
Oflag IXA, Spangenberg castle, Kassel, 71, 77-79
Oflag IXA/H, 77-78, 79
Oflag XXIB, Schubin, Poland, 71
Olson, Virgil, 4, 7, 13
Operation *Jubilee*, 75, 77
Operation *Sealion*, Hitler's invasion of Britain, 2
Orbinson, Edwin "Bud", 9, 12-13, 15
Osborne, Julian, 75-76
Osbourne, Arthur, 105

P

Pacific Theatre, 45-51
Palembang, 46-48
Paris, France, German occupation, 73-74
Parker, Vernon A. "Shine", 34-42, 71
Patterson, Robert G. "Pat", 113-117, 181
Patton, George, 180
Pearl Harbor, 24, 43
Peck, E.E., 39, 40
Peel, Argenta, 36
Peel, Doriel, 36
Peel, John and Joan, 36
Peel, W.E., 36
Peters, Alex, 38, 76
Peterson, Chesley, 9, 11, 15-16, 27, 32, 75, 151
Peterson, Kenneth, 120
Pisanos, Steve, 120, 148-154
Poland, German invasion of, 1
Priller, Joseph, 107
Priser, Robert L., 75, 76, 143-148

R

Resistance, French (see Free French Underground)
Reston, Scott, 7
Reynolds, Quentin, 9-11
Robinson, J. Roland "Robbie" Lord Martonmere, 12, 33
Rommel, Erwin, 57
Ross, Donald H., 117
Ryder, Frances, 35
Ryerson, Leonard T., 83-86

S

Saturday Evening Post, 10
Scanlan, Michael, 7

Scharff, Hanns, 132, 133
Scotland, 36-42
Sealion, Hitler's invasion of Britain, 2
Shubin XXIB, 104
Silvers, Lennie Marting, 68
Sinclair, Archibald, 8
Singapore, 45
Skinner, Leroy, 72, 98, 100-103, 166, 167
Smith, Bradley, 52
Smith, Dennis D., 83-86
Smith, Fonzo D. "Don/Snuffy", 120, 134-141, 162-164, 182
Smith, Kenneth G., 120, 138
Smith, Robert E. "Bob", 90-95
Spaatz, Carl "Tooey", 87
Spangenberg (*see* Oflag IXA)
Spanish Civil War, Lincoln Brigade, 2
Sperry, George B., 81, 84-88, 96-97, 111, 139, 142, 162, 164, 165, 180, 181
Spicer, Russ, 119
Spitfire aircraft, 6, 13, 82
St. Nazaire, 5
St. Omer hospital, France, 17-19, 23
Stalag Luft I, Barth, Germany, 113, 117, 119, 134, 141-142
Stalag Luft III, Sagan, Germany, 16, 19, 71, 72, 77, 79, 81, 84, 95-112, 126, 132, 138-139, 161
Stalag Luft VIIA, Moosburg, Bavaria, 164-166, 176, 180-183
Stalag Luft VIID, 173
Stanhope, Aubrey, 108
Statroda POW hospital, 20
Stettin raid, 129
Stillman, R.M., 99, 110
Stolle, Bruno, 84, 86
Stout, Roy, 13
Stukas, Joseph, 117
Sweeny, Charles, 8
Sweeny, Robert, 8, 9, 12

T

Tally Ho! Yankee In a Spitfire, 47
Taylor, Edwin D. "Jessie", 76
Taylor, James, 76
Taylor, William E.G., 8, 11

Thorpe, Cliff, 52, 56
Thunderbolt aircraft, 105
Tiger Moth aircraft, 39
Tobin, Eugene "Red," 3-15, 34
Tobin, I. Quimby, 3-4
Toft, Kenneth, 104
Trenthorst, 181
Tribken, Wally, 23, 56, 69
Tumult In The Clouds, 131
Typhoon fighters, 75

U

Underground, French (*see* Free French Underground)
uniform of Eagle Squadron, 8

V

Vibert, John A., 47-48
Vichy France (*see* Free French Underground)

W

Walsh, Paddy, 41
War Eagles, 13
Warmwell, Dorset, England, 7
White, Bill, 13
Whitlow, Gordon, 105, 107
Wiendels, Aren, 114
Wiener-Neustadt raid, 121-126
Wilkinson, Royce C. "Wilkie", 72, 73, 75
Willis, Donald K. "Don", 154-160
Willkie, Wendell, 12
Winant, John, 151
Windom, Rhea, 39
Witt, Harry, 10
Wolfe, Roland E. "Bud", 24-25
Woodhouse, Henry "Paddy", 17, 27
Wilkinson, Royce, 9
WRENS, 44
Wright, Gil, 111

X

X Committee, Stalag Luft III, 98, 100
Xenia, Paul, 42

Y

Young, Norman, 76